Biochar - An Imperative Amendment for Soil and the Environment

Edited by Vikas Abrol and Peeyush Sharma

Published in London, United Kingdom

IntechOpen

Supporting open minds since 2005

Biochar - An Imperative Amendment for Soil and the Environment
http://dx.doi.org/10.5772/intechopen.74890
Edited by Vikas Abrol and Peeyush Sharma

Contributors
Kayode Are, Hidetaka Noritomi, Jyoti Saxena, Jyoti Rawat, Pankaj Sanwal, Thavivongse Sriburi, Saowanee Wijitkosum, Guangwei Yu, Shengyu Xie, Futian You, Héctor U. Levatti, Jie Li, Chunxing Li, Xiaoda Tang, Lanjia Pan, Yin Wang, Xiaofu Shang, Cheng Yu, Jianli Ma, Lucy Ngatia, Johnny Grace III, Robert Taylor, Daniel Moriasi, George Osei, Alejandro Bolques

First published in London, United Kingdom, 2019 by IntechOpen
IntechOpen is the global imprint of INTECHOPEN LIMITED, registered in England and Wales, registration number: 11086078, The Shard, 25th floor, 32 London Bridge Street
London, SE19SG - United Kingdom
Printed in Croatia

British Library Cataloguing-in-Publication Data
A catalogue record for this book is available from the British Library

Additional hard and PDF copies can be obtained from orders@intechopen.com

Biochar - An Imperative Amendment for Soil and the Environment
Edited by Vikas Abrol and Peeyush Sharma
p. cm.
Print ISBN 978-1-83881-987-3
Online ISBN 978-1-83881-988-0
eBook (PDF) ISBN 978-1-83881-989-7

For EU product safety concerns:
IN TECH d.o.o., Prolaz Marije Krucifikse Kozulić 3, 51000 Rijeka, Croatia, info@intechopen.com or visit our website at intechopen.com.

Meet the editors

Dr. Vikas Abrol (b. November 14, 1974), a soil scientist, received his PhD degree from Sher-e-Kashmir University of Agricultural Sciences and Technology-Jammu and was awarded a postdoctoral research fellowship for one year by the Ministry of Agriculture, State of Israel, to pursue research into biochar: a mechanism of action as a soil and water conservation agent in agricultural soils at the Volcani Centre, Agricultural Research Organization, Israel. He also received a research fellowship for one academic year at the Volcani Centre, Agricultural Research Organization, Israel, to pursue research on the efficacy of synthetic polymers (anionic polyacrylamide (PAM)) on soil and water conservation. He started his professional career as an assistant professor/junior scientist at the Dryland Research Substation, Dhiansar, and has published research accomplishments in journals of international repute such as the *European Journal of Agronomy*, *European Journal of Soil Science*, *Journal of Soil and Sediments*, *Journal of the Science of Food and Agriculture*, *Agricultural Mechanization in Asia*, *Africa and Latin America*, etc. with high impact factor. He has edited two international books: *Crop Production Technologies* and *Resource Management for Sustainable Agriculture*. He served as a reviewer of research articles in international journals such as *Soil Science Society of America Journal*, *Agronomy for Sustainable Development*, *Land Research and Development*, *Indian Journal of Agricultural Sciences*, *Indian Journal of Dryland Agricultural Research and Development*, *Indian Journal of Soil Conservation*, *Journal of Experimental Biology and Agricultural Sciences*, and *International Journal of Agriculture Sciences*. His research interests entail using biochar for offsetting climate change by soil carbon aggradations, runoff quality and soil erosion control, soil quality evaluation, and soil and water pollution. He holds specialization in using biochar and synthetic polymers (PAM) for erosion control, infiltration improvement, and soil structural stabilization. He also served in Krishi Vigyan Kendra (extension services) and as a resource person for Prasar Bharti, Department of Agriculture, SAMETI, ATMA, National Fertilizer Limited, Indian Potash Association. He has presented papers in national and international conferences/seminars and has been awarded in national and international forums for his scientific contributions.
He is a life member of the *Indian Journal of Soil Conservation*, *Indian Journal of Dryland Agriculture and Research*, *International Biochar Initiative*, and *Friends of Israel Biochar Researchers Network*.

Dr. Peeyush Sharma received her MSc and PhD (Soil Science) degrees from the illustrious G.B. Pant University of Agriculture and Technology, Pantnagar, Uttaranchal, and was awarded a postdoctoral fellowship on the "Development and validation of simulation models to predict the long-term consequence of different tillage and residue management strategies." Later she was also awarded a postdoctoral fellowship from the prestigious Institute of Soil Water and Environmental Sciences, ARO, Volcani Centre, Israel, and joined Sher-e-Kashmir University of Agricultural Sciences and Technology-Jammu in 2004. Primarily, her research pertains to modeling tillage with the intervention of mulching and nutrient management to aggrade soil health and develop a "Tilth Index Model" based on easily measurable soil properties to predict changes in soil productivity and identify the suitable tillage type needed to achieve an optimal crop production. She holds expertise in low-cost biochar production technology and application in the soils of diverse agroclimatic regions of the Himalayan foothills. Her teaching assignment involves advances in soil physics, fundamentals of soil science, and other courses for UG and PG. Her research accomplishments include contributions to reputed international journals such as *Geoderma*, *European Journal of Agronomy*, *European Journal of Soil Science*, and *Soil and Tillage Research* and other reputed journals, and she has served as a reviewer for *Agriculture and Water Management*, *African Journal of Agricultural Sciences*, *Indian Journal of Agricultural Sciences*, *Indian Journal of Soil Conservation*, etc.

Contents

Preface

Climate change, water scarcity, soil degradation, and greenhouse gas (GHG) emission are the major challenges for food security. The world's population is expected to reach 9 billion by 2050, which will require an increase of more than 50% in agricultural food supply to meet the growing demand. Throughout the world, agricultural crop residue is usually handled as a liability, often because the means to transform it into an asset are lacking. Concerns regarding global warming and food security have led to a surge in interest in the management of crop residues to increase carbon sequestration and grain yield in agriculture. Crop residue burning has become a major environmental problem causing health issues as well as contributing to global warming and the degradation in soil health. India, being an agriculture-dominant country and the second largest agro-based economy with year-round crop cultivation, produces a large amount of agricultural waste. According to the Indian Ministry of New and Renewable Energy, India generates on an average 500 million tons of crop residue per year. In the absence of adequate sustainable management practices, approximately 92 metric tons of crop waste are burned every year in India, causing excessive particulate matter emissions and air pollution. Biochar offers the opportunity to provide a sustainable solution to mitigate these issues.

Soil organic carbon (SOC) content, which plays an important role in soil sustainability, is a key indicator of soil fertility. SOC is the basis of soil fertility. It releases nutrients for plant growth, promotes the structural, biological, and physical health of soil, and is a buffer against harmful substances. The main source of SOC in cropland is crop residue; therefore, crop residue amendment is considered one of the most important management practices in maintaining soil fertility. Efficient use of biomass by converting it to a useful source of soil amendment is one way to improve soil fertility. Agriculture is a major source of GHG emissions globally. Increased use of production inputs, such as mineral fertilizer, has made agriculture more GHG intensive. Rising concentration of the GHG carbon dioxide in the atmosphere is a major anthropogenic cause of climate change. The changing climate impacts society and ecosystems in many harmful ways.

To increase carbon sequestration, farmers can use biochar, which is the pyrolyzed product of the thermal degradation of organic materials in the absence of oxygen and is distinguished from charcoal by its use as a soil amendment. Over the past few years, pyrolyzed organic carbon has received much attention from researchers because of the possible benefits arising from soil quality and crop yields. It is a unique substance that retains exchangeable and plant-available nutrients in the soil, improving crop yields while decreasing environmental pollution by nutrients. Biochar is an effective means to withdraw carbon dioxide from the atmosphere and consequently influence the trend of global climate change. The physical structure of biochar improves soil fertility by decreasing soil tensile strength and bulk density. Biochar also acts as a soil aggregate, which provides a habitat for microorganisms. Additionally, the porosity of biochar improves nutrient and water retention in soils thereby improving agronomic efficiency and increasing yields. It has often been referred to, not only as one of the possible means for enhancing soil fertility, but as black gold for agriculture as well. If the fertility or nutrient status of soil is increased, it leads to an increase in crop production and plays a significant role in maintaining soil quality and health. It offers not only an attractive solution for reducing air pollution from the open burning of crop residues, but is also a favorable sustainable model for reutilizing agricultural wastes.

This book attempts to gather and discuss the information and technologies developed for biochar production and its application to agriculture. The emphasis in this endeavor is on the use of biochar in agriculture for improving soil health, crop productivity, and GHG mitigation. This book contains chapters that look at outcomes of biochar research being conducted in different parts of India, and the potential benefits of biochar use in improving soil health, crop productivity, and in mitigating climate change through reduction in emission of GHGs and carbon sequestration. Biochar has great potential for improving soil fertility and crop productivity.

We are thankful to the authors who are experts in their respective fields, and who have written a comprehensive and valuable resource for researchers, academicians, and students interested in gaining knowledge of role of biochar in the field of agricultural sustainability.

Last but not the least, we acknowledge wholeheartedly IntechOpen for publishing this book for the benefit of the agricultural fraternity.

Vikas Abrol and Peeyush Sharma
Division of Soil Science and Agricultural Chemistry,
Sher-e-Kashmir University of Agricultural Sciences and Technology,
India

Section 1

Biochar for Sustainable Crop Productivity and Soil Health

Chapter 1

Biochar: A Sustainable Approach for Improving Plant Growth and Soil Properties

Jyoti Rawat, Jyoti Saxena and Pankaj Sanwal

Abstract

Soil is the most important source and an abode for many nutrients and microflora. Due to rapid depletion of agricultural areas and soil quality by means of ever-increasing population and an excessive addition of chemical fertilizers, a rehabilitated attention is a need of the hour to maintain sustainable approaches in agricultural crop production. Biochar is the solid, carbon-rich material obtained by pyrolysis using different biomasses. It has been widely documented in previous studies that, the crop growth and yield can be increased by using biochar. This chapter exclusively summarizes the properties of biochar, its interaction with soil microflora, and its role in plant growth promotion when added to the soil.

Keywords: biochar, pyrolysis, soil microflora, nutrients, plant growth promotion

1. Introduction

Crop growth and productivity are strongly influenced by various biotic and abiotic stresses such as pests, weeds, drought, high salinity, extreme temperature, etc. and the soil quality [1]. Soil is also contaminated by heavy metals through various human activities [2], which affect plant growth and development and ultimately brings low yielding cropping systems. Mining is one of the important sources of heavy metal contamination in soil [3, 4]. The strength of soil is directly related to nutrient availability. Plants require a number of soil nutrients like nitrogen (N), phosphorus (P), and potassium (K) for their growth, but soil nutrient levels may decrease over time after crop harvesting, as nutrients are not returned to the soil. In India, the soil of many regions is not only deficient in macronutrients like NPK but also in secondary nutrients (e.g. sulfur, calcium, and magnesium) and micronutrients (e.g. boron, zinc, copper, and iron) [5]. Thus, to fulfill the shortage, a large amount of chemical fertilizers is added to the soil; however, only a small percent of water-soluble nutrients are taken up by the plants and the rest are converted into insoluble forms, making continuous application necessary. Finally, the extensive use of chemical fertilizers has led to the deterioration of the environment causing infinite problems. It not only lowers the nutrient composition of the crops but also degrades the soil fertility in the long run [6, 7].

Besides fertilizers, pesticides are also the basic evil for agriculture, and the adverse effects of pesticides on the environment are truly responsible for influencing the microbial properties of soil. High inputs of fertilizers and pesticides and their long persistence in the soil adversely affect the soil microflora, thereby disturbing soil health and significantly reducing the total bacterial and fungal biomass [8]. Due to long-term treatment with inorganic fertilizers (N and NPK) and/or organic manures, a shift in structural diversity and dominant bacterial groups in agricultural soils has been recorded by Wu et al. [9]. Biofertilizers, on the other hand, can reenergize the soil by improving the soil fertility and hence can be used as a powerful tool for sustainable agriculture, rendering agro-ecosystems more stress-free. Additionally, the application of organic amendments to soils, from a remedial point of view, has typically been justified by their relatively low cost, which normally requires other forms of disposal (burial in a landfill, incineration, etc.). Soil amendments must possess properties such as high binding capacity and environmental safety and should have no negative effect on the soil structure, soil fertility, or the ecosystem on the whole [10]. The use of biochar has been accepted as a sustainable approach and a promising way to improve soil quality and remove heavy-metal pollutants from the soil [11].

Biochar is a carbon-rich organic material, an organic amendment, and a by-product derived from biomass by pyrolysis under high-temperature and low-oxygen conditions. Biochar is produced through a process called pyrolysis, which basically involves heating of biomass (such as wood, manure, or leaves) in complete or almost complete absence of oxygen, with oil and gas as co-products. However, the quantity of these materials produced depends on the processing conditions. Recently, it has been reported that biochar obtained from the carbonization of organic wastes can be a substitute that not only influences the sequestration of soil carbon but also modifies its physicochemical and biological properties [12, 13].

Biochar has the potential to produce farm-based renewable energy in an eco-friendly way. Specifically, the quality of biochar depends on several factors, such as the type of soil, metal, and the raw material used for carbonization, the pyrolysis conditions, and the amount of biochar applied to the soil [14]. In addition, the biochar amendment to the soil proved to be beneficial to improve soil quality and retain nutrients, thereby enhancing plant growth [15]. Since biochar contains organic matter and nutrients, its addition increased soil pH, electric conductivity (EC), organic carbon (C), total nitrogen (TN), available phosphorus (P), and the cation-exchange capacity (CEC) [16]. Earlier, Verheijen et al. [17] reported that the biochar application affected the toxicity, transport, and fate of various heavy metals in the soil due to improved soil absorption capacity. The presence of plant nutrients and ash in the biochar and its large surface area, porous nature, and the ability to act as a medium for microorganisms have been identified as the main reasons for the improvement in soil properties and increase in the absorption of nutrients by plants in soils treated with biochar [18]. Chan et al. [19] reported that biochar application decreased the tensile strength of soil cores, indicating that the use of biochar can reduce the risk of soil compaction. A lot has already been discussed on the benefits of inoculation of rhizobacteria in soil, but the addition of biochar can also provide more nutrients to the soil, thus benefiting the agricultural crops. The mixing of the plant growth-promoting microorganisms with biochar was referred to as the best combination for growth and yield of French beans by Saxena et al. [20].

Addition of biochar in the soil can be extremely useful to improve the soil quality, as well as to stimulate the plant growth, and thus, biochar can play an important role in developing a sustainable system of agriculture. Several uses and positive effects of biochar amendment have currently been considered as an effective method to reclaim the contaminated soil [21] and to achieve high crop yields without harming the natural environment. The positive influence of biochar on plant growth and soil quality suggests that using biochar is a good way to overcome nutrient deficiency, making it a suitable technique to improve farm-scale nutrient cycles. Therefore, a complete focus is been made to explore the positive effects of biochar amendment on soil stability and plant growth promotion.

2. Biochar production and properties

Biochar is made up of elements such as carbon, hydrogen, sulfur, oxygen, and nitrogen as well as minerals in the ash fraction. It is produced during pyrolysis, a thermal decomposition of biomass in an oxygen-limited environment. Biochar is black, highly porous, and finely grained, with light weight, large surface area and pH, all of which have a positive effect on its application to soil. To address the major concern on quality of agricultural soil degradation, biochar is applied to the soil in order to enhance its quality. Biochar is stabilized biomass, which may be mixed into soil with intentional changes in the properties of the soil's atmosphere to increase crop productivity and to mitigate pollution. The raw material (biomass) used and processing parameters dictate the properties of the biochar.

2.1 Biomass as a raw material

A wide range of organic materials are suitable as feedstock for the production of biochar. Biochar can be produced with raw materials such as grass, cow manure, wood chips, rice husk, wheat straw, cassava rhizome, and other agricultural residues [22, 23]. It was reported that the production of biochar with high nutrients depends on the type of raw material used and pyrolysis conditions [24]. Biochar is produced from the residual biomasses such as crop residues, manure, wood residues, and forests and green wastes using modern pyrolysis technology. Agricultural wastes (bark, straw, husks, seeds, peels, bagasse, sawdust, nutshells, wood shavings, animal beds, corn cobs and corn stalks, etc.), industrial wastes (bagasse, distillers' grain, etc.), and urban/municipal wastes [25, 26] have been extensively used, thus also achieving waste management through its production and use [27].

Feedstocks currently used on a commercial scale include tree bark, wood chips, crop residues (nut shells, straw, and rice hulls), grass, and organic wastes including distillers' grain, bagasse from the sugarcane industry, mill waste, chicken litter, dairy manure, sewage sludge, and paper sludge [28–30]. A 40 wt.% yield of biochar from maize stover was obtained by Peterson et al. [31].

The biomass used for the production of biochar is mainly composed of cellulose, hemicellulose, and lignin polymers [32]. Among these, cellulose has been found to be the main component of most plant-derived biomasses, but lignin is also important in woody biomass.

2.2 Biochar production

Biochar can be manufactured on a small scale using low-cost modified stoves or kilns or through large-scale, cost-intensive production, which utilizes larger pyrolysis plants and higher amounts of feedstocks. Biochar is produced from several biomass feedstocks through pyrolysis as discussed above, generating oil and gases as by-products [33]. The dry waste obtained is simply cut into small pieces to less than 3 cm prior to use. The feedstock is heated either without oxygen or with little oxygen at the temperatures of 350–700°C (662–1292°F). Pyrolysis is generally classified by the temperature and time duration for heating; fast pyrolysis takes place at temperatures above 500°C and typically happens on the order of seconds (heating rates $\geq$ 1000°C/min). This condition maximizes the generation of bio-oil. Slow pyrolysis, on the other hand, usually takes more time, from 30 min to a few hours for the feedstock to fully pyrolyze (heating rates $\leq$ 100°C/min) and at the same time yields more biochar. The temperature range remains 250–500°C [34].

The type of biochar produced depends on two variables: the biomass being used and the temperature and rate of heating. High and low temperatures have an unequivocal effect on char yields. It has been noticed that at low temperature (<550°C), biochar has an amorphous carbon structure with a lower aromaticity than the biochar produced at high temperature [35]. High temperature leads to lower char yield in all pyrolysis reactions [36]. Peng et al. [37] reported the effect of charring duration on the yield of biochar; yield showing a decrease with increasing duration at the same temperature. The pyrolysis process seriously affects the quality of biochar and its potential value to agriculture in terms of agronomic performance or in carbon sequestration. The yield of biochar from slow pyrolysis of biomass has been stated to be in the range of 24–77% [38, 39] (**Figure 1**). The pyrolysis process can be shown as follows:

$$\text{Biomass (Solid)} \rightarrow \text{Biochar} + \text{Liquid or oil (tars, water, etc.)} + \text{Volatile gases } (CO_2, CO, H_2) \quad (1)$$

2.3 Physical, chemical and biological properties of biochar

Biochar is a stable form of carbon and can last for thousands of years in the soil [40]. It is produced for the purpose of addition to soil as a means of sequestering carbon and improving soil quality. The conditions of pyrolysis and the materials used can significantly affect the properties of biochar. The physical properties of biochar contribute to its function as a tool for managing the environment. It has been reported that when biochar is used as a soil amendment, it stimulates soil fertility and improves soil quality by increasing soil pH, increasing the ability to retain moisture, attracting more useful fungi and other microbes, improving the ability of

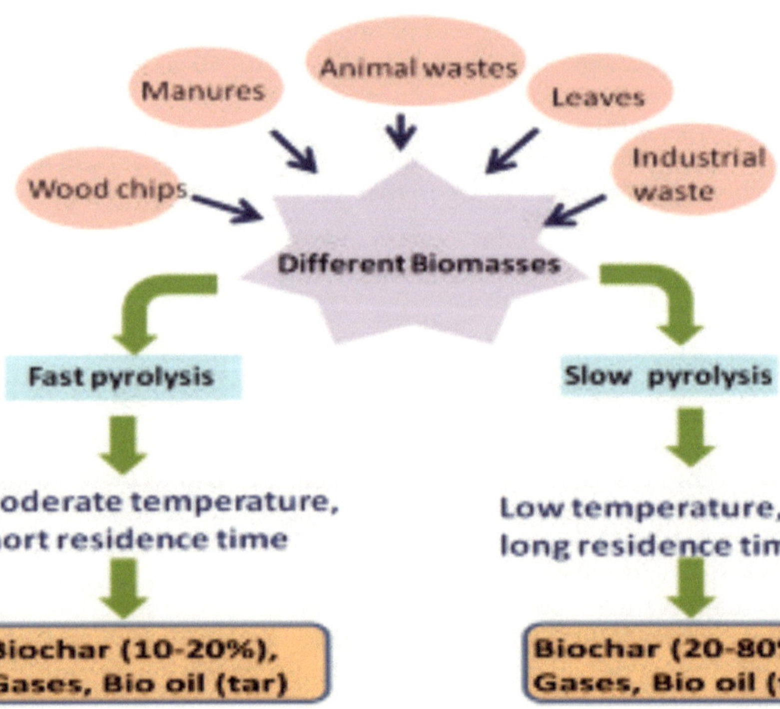

Figure 1.
Biochar production from different biomasses.

cation exchange, and preserving the nutrients in the soil [41]. Biochar reduces soil density and soil hardening, increases soil aeration and cation-exchange capacity, and changes the soil structure and consistency through the changes in physical and chemical properties. It also helps to reclaim degraded soils. It has shown a greater ability to adsorb cations per unit carbon as compared to other soil organic matters because of its greater surface area, negative surface charge, and charge density [42], thereby offering the possibility of improving yields [43]. Samples with a sufficient amount of stable carbon can be added to the soil to be sequestered; a high sorption surface of biochar can characterize it as a soil additive, competent of halting risk elements in soil.

The physical characteristics of biochar are directly and indirectly related to how they affect soil systems. Soils have their own physical properties depending on the nature of mineral and organic matter, their relative amounts, and how minerals and organic matter are related. When biochar is present in the soil mixture, its contribution to the physical nature of the system is significant, affecting the depth, texture, structure, porosity, and consistency by changing the surface area, pore and particle-size distribution, density, and packing [44]. The influence of biochar on physical properties of soil directly affects the growth of plants, since the depth of penetration and accessibility of air and water in the root zone is determined mainly by the physical composition of the soil horizons. This affects the soil's response to water, its aggregation, and work ability in soil preparation, dynamics, and permeability when swelling, as well as the ability to retain cations and response to changes at ambient temperature. The smaller the pores on biochar, the longer they can retain capillary soil water. The addition of biochar can reduce the effects of drought on crop productivity in drought-affected areas due to its moisture-retention capacity. It has been shown that it eliminates soil constraints that limit the growth of plants, and neutralizes acidic soil because of its basic nature [45]. Carbon dioxide and oxygen occupy air-filled spaces on the pores of biochar or can be chemosorbed on the surface. As biochar can contain nutrients, microorganisms, and syngases, it can also retain fertilizers in the soil longer than other soils and prevent it from leaching into water sources such as rivers and lakes.

As far as its chemical properties are concerned, biochar reduces soil acidity by increasing the pH (also called the liming effect) and helps the soil to retain nutrients and fertilizers [46]. The application of biochar improves soil fertility through two mechanisms: adding nutrients to the soil (such as K, to a limited extent P, and many micronutrients) or retaining nutrients from other sources, including nutrients from the soil itself. However, the main advantage is to retain nutrients from other sources. In most cases, the addition of biochar only has a net positive effect on the growth of crops if nutrients from other sources, such as inorganic or organic fertilizers, are used. Biochar increases the availability of C, N, Ca, Mg, K, and P to plants, because biochar absorbs and slowly releases fertilizers [47]. It also helps to prevent fertilizer drainage and leaching by allowing less fertilizer use and reducing agricultural pollution in the surrounding environment [48]. Biochar alleviates the impact of hazardous pesticides and complex nitrogen fertilizers from the soil, thus reducing the impact on the local environment.

Good healthy soil should include a wide and balanced variety of life forms, including bacteria, fungi, protozoa, nematodes, arthropods, and earthworms. Recently, biochar has been reported to increase the microbial respiration of the soil by creating space for soil microbes [49], and in turn the soil biodiversity and soil density increased. Biochar also served as a habitat for extra-radical fungal hyphae that sporulated in micropores due to lower competition from saprophytes and therefore served as an inoculum for arbuscular mycorrhizal fungi [50]. It is believed that biochar has a long average dwelling time in soil, ranging from 1000 to 10,000 years, with an average of 5000 years [51–53]. However, its recalcitrance and physical nature present significant impediment to the evaluation of long-term stability [43]. The commercially available soil microbes which can be used for inoculation include *Azospirillum* sp., *Azotobacter* sp., *Bacillus thuringiensis*, *B. megaterium*, *Glomus fasciculatum*, *G. mosseae*, *Pseudomonas fluorescens*, *Rhizobium* sp., and *Trichoderma viride* [54].

3. Biochar as a soil amendment

The issues as food security, declining soil fertility, climate change, and profitability are the driving forces behind the introduction of new technologies or new farming systems. The amendment of soils for their remediation aims at reducing the risk of pollutant transfer to waters or receptor organisms in proximity. The organic material such as biochar may serve as a popular choice for this purpose because its source is biological and it may be directly applied to soils with little pretreatment [55]. There are two aspects which make biochar amendment superior to other organic materials: the first is the high stability against decay, so that it can remain in soil for longer times providing long-term benefits to soil and the second is having more capability to retain the nutrients. Biochar amendment improves soil quality by increasing soil pH, moisture-holding capacity, cation-exchange capacity, and microbial flora [56].

The addition of biochar to the soil has shown the increase in availability of basic cations as well as in concentrations of phosphorus and total nitrogen [57, 58]. Typically, alkaline pH and mineral constituents of biochar (ash content, including N, P, K, and trace elements) can provide important agronomic benefits to many soils, at least in the short to medium term. When biochar with a higher pH value was applied to the soil, the amended soil generally became less acidic [59]. Acidic biochar could also increase soil pH when used in soil with a lower pH value. The pH of biochar, similar to the other properties, is influenced by the type of feedstock, production temperature, and production duration.

Another valuable property of biochar is suppression of emissions of greenhouse gases in soil. It has also been demonstrated by Zhang et al. [60] that the emissions of methane and nitrous oxide were reduced from agricultural soils, which may have additional climate mitigation effects, since these are potent greenhouse gases. Spokas et al. [61] reported reduced carbon dioxide production by addition of different concentrations of biochar ranging from 2 to 60% (w/w), suppressed nitrous oxide production at levels higher than 20% (w/w), and ambient methane oxidation at all levels over unamended soil.

Several studies have shown the control of pathogens by the use of biochar in agricultural soil. Bonanomi et al. [62] reported that biochar is effective against both air-borne (e.g. *Botrytis cinerea* and different species of powdery mildew) and soil-borne pathogens (e.g. *Rhizoctonia solani* and species of *Fusarium* and *Phytophthora*). The application of the biochar derived from citrus wood was capable of controlling air-borne gray mold, *Botrytis cinerea* on *Lycopersicon esculentum*, *Capsicum annuum* and *Fragaria × ananassa*. Although there is a shortage of published data on the effects of biochar on soil-borne pathogens, evidence given by Elmer et al. [63] has shown that the control of certain pathogens may be possible. The addition of biochar in 0.32, 1.60, and 3.20% (w/w) to asparagus soils infested with *Fusarium* has augmented the biomass of asparagus plants and reduced *Fusarium* root rot disease [63]. Similarly, *Fusarium* root rot disease in asparagus was also reduced by biochar inoculated with mycorrhizal fungi [64]. A study of suppression of bacterial wilt in tomatoes showed that biochar obtained from municipal organic waste reduced the incidence of the disease in *Ralstonia solanacearum* infested soil [65]. Ogawa [66] advocated the use of biochars and biochar amended composts for controlling the diseases caused by bacteria and fungi in soil. The disease suppression mechanism has been attributed to the presence of calcium compounds, as well as improvements in the physical, chemical, and biological characteristics of the soil.

The prevention of 'diffuse water pollution' through ammonium sorption or the mediation of the dynamics of a soil solution containing nitrate, phosphorus, and other nutrients has been extensively studied. The application of biochar to soil can influence a wide range of soil constraints such as high availability of Al [67], soil structure and nutrient availability [24], bioavailability of organic [68] and inorganic pollutants [69], cation-exchange capacity (CEC), and retention of nutrients [70, 71]. Biochar can also adsorb pesticides, nutrients, and minerals in the soil, preventing the movement of these chemicals into surface water or groundwater and the subsequent degradation of these waters from agricultural activity.

Xie et al. [72] reported that biochar amendment enhanced soil fertility and crop production, particularly in soils with low nutrients. However, in soils with high fertility, no noticeable increase in production was noticed, and some studies even reported inhibition of plant growth. The observations of Taghizadeh-Toosi et al. [73] indicated that ammonia adsorbed by biochar could be later released to the soil. Saarnio et al. [74] showed that biochar application along with fertilizers can lead to better plant growth, but sometimes a negative effect was also observed without fertilization due to reduced bio-availability through sorption of nitrogen. It has been shown that application of biochar in the soil has a positive to neutral and even negative impact on crop production. Hence, it is crucial that the mechanisms for action of biochar in the soil be understood before its application.

The consequence of biochar addition on plant productivity depends on the amount added. Recommended application rates for any soil amendment should be based on extensive field testing. At present, insufficient data are available for obtaining general recommendations. In addition, biochar materials can vary greatly in their characteristics, so the nature of the particular biochar material (e.g. pH and ash content) also influences the application rate. Several studies have reported a positive effect of using biochar on crop yields with rates of 5–50 tonnes per hectare with appropriate nutrient management. The experiments conducted by Rondon et al. [75] resulted in a decrease in crop yield in a pot experiment with nutrient deficient soil amended with biochar at the rate of 165 tonnes per hectare. An experiment conducted in the United States showed that peanut hull and pine chip biochar, applied to 11 and 22 tonnes per hectare, could reduce corn yields below those obtained in the control plots with standard fertilizer management [76]. Thus, the control of the rate of application of biochar is necessary to prevent the negative impact of biochar.

4. Stimulation of soil microflora and plant growth

There are several reports which show that biochar has the capability to stimulate the soil microflora, which results in greater accumulation of carbon in soil. Besides adsorbing organic substances, nutrients, and gases, biochars are likely to offer a habitat for bacteria, actinomycetes and fungi [64]. It has been suggested that faster heating of biomass (fast pyrolysis) will lead to the formation of biochar with fewer microorganisms, smaller pore size, and more liquid and gas components [77]. The enhancement of water retention after biochar application in soil has been well established [78], and this may affect the soil microbial populations. Biochar provides a suitable habitat for a large and diverse group of soil microorganisms, although the interaction of biochar with soil microorganisms is a complex phenomenon. Many studies reported that addition of biochar along with phosphate solubilizing fungal strains promoted growth and yield of *Vigna radiata* and *Glycine max* plants, with better performances than control or those observed when the strains and biochar are used separately [20, 79, 80].

The use of biochar increased mycorrhizal growth in clover bioassay plants by providing the suitable conditions for colonization of plant roots [81]. Warnock et al. [82] summarized four mechanisms by which biochar can affect functioning of mycorrhizal fungi: (i) changes in the physical and chemical properties of soil, (ii) indirect effects on mycorrhizae through exposure to other soil microbes, (iii) plant-fungus signaling interference and detoxification of toxic chemicals on biochar, and (iv) providing shelter from mushroom browsers. Carrots and legumes grown on steep slopes and in soils with less than 5.2 pH showed significantly improved growth by the addition of biochar [83]. It was found that biochar increased the biological N_2 fixation (BNF) of *Phaseolus vulgaris* [75] mainly due to greater availability of micronutrients after application of biochar. Lehmann et al. [58] reported that biochar reduced leaching of NH_4^+ by supporting it in the surface soil where it was available for plant uptake. Mycorrhizal fungi were often included in crop management strategies as they were widely used as supplements for soil inoculum [84]. When using both biochar and mycorrhizal fungi in accordance with management practices, it is obviously possible to use potential synergism that can positively affect soil quality. The fungal hyphae and bacteria that colonize the biochar particles (or other porous materials) may be protected from soil predators such as mites, Collembola and larger (>16 μm in diameter) protozoans and nematodes [85–87].

Biochar can increase the value of non-harvested agricultural products [88] and promote the plant growth [58, 89]. A single application of 20 t ha^{-1} biochar to a Colombian savanna soil resulted in an increase in maize yield by 28–140% as compared with the unamended control in the 2nd to 4th years after application [90]. With the addition of biochar at the rate of 90 g kg^{-1} to tropical, low-fertile ferralsol, not only the proportion of N fixed by bean plants (*Phaseolus vulgaris*) increased from 50% (without biochar) to 72%, but also the production of biomass and bean yield were improved significantly [75]. When biochar was applied to the soil, a higher grain yield of upland rice (*Oryza sativa*) was obtained in northern Laos sites with low P availability [91, 92]. Many of these effects are interrelated and may act synergistically to improve crop productivity. Often there has been a reported increase in yields, which is directly related to the addition of biochar as compared to the control (without biochar) [58]. However, in some cases, growth was found to be depressed [93].

The direct beneficial effects of biochar addition for the availability of nutrients are largely due to the higher content of potassium, phosphorus, and zinc availability and, to a lesser extent, calcium and copper [58]. Few studies have examined the potential for amending biochar in soil to impact plant resistance to pathogens. With reference to soil pathogens principally concerned with the effect of AM fungal inoculations on asparagus tolerance to the soil borne root rot pathogen *Fusarium*, Matsubara et al. [94] demonstrated that charcoal amendments had a suppressive effect on pathogens. One more study that supported these earlier findings stated that biochar made from ground hardwood added to asparagus field soil led to a decrease in root lesions caused by *Fusarium oxysporum*, *F. asparagi*, and *F. proliferatum* compared to the non-amended control [95]. Biochar reduces the need for fertilizer, which results in reduction in emissions from fertilizer production, and turning the agricultural waste into biochar also reduces the level of methane (another potent greenhouse gas) caused by the natural decomposition of waste.

5. Mixing biochar with other amendments

Mixing biochar with other soil amendments such as manure, compost, or lime before soil application can improve efficiency by reducing the number of field operations required. Since biochar has been shown to sorb nutrients and protect them from leaching [70, 96], mixing of biochar may improve the efficiency of manure and other amendments. However, Kammann et al. [97] acknowledged in their recent review that very few studies that directly combined organic amendments with biochars were available. They found that co-composted biochars had a remarkable plant growth-promoting effect as compared to biochars when used pure, but no-systematic studies have been done to understand the interactive effects of biochars with non-pyrogenic organic amendments (NPOAs). Biochar can also be mixed with liquid manures and used as slurry. Additionally, combined biochar and compost applications have numerous advantages over mixing of biochar or compost with soil separately. These benefits, according to Liu et al. [98], include more efficient use of nutrients, biological activation of biochar, an enhanced supply of plant-available nutrients by biological nitrogen fixation, reduction of nutrient leaching, and the contribution of combined nutrients in comparison to a single application of compost and biochar. Diminutive biochars are most likely best suited for this type of application. Biochar was also mixed with manure in ponds and potentially reduced losses of nitrogen gas were recorded same as when it was applied to soil [99, 100].

6. Conclusion

The problem of the depletion of agricultural land as a result of the pressure caused by the ever-growing population necessitated the sustainable practice of crop production. It was suggested to use biochar as a means of remediating contaminated agricultural soil, improving soil fertility by reducing the acidity, and increasing the availability of nutrients. Thus, addition of biochar to the soil can be one of the best practices to overcome any biotic stress in soil and to increase the crop productivity. The positive effects of biochar on the interactions between soil-plant-water caused better photosynthetic performance and improved nitrogen and water use efficiency. Hence, it can be concluded by this comprehensive review that biochar has the potential to improve the properties of soil, microbial abundance, biological nitrogen fixation, and plant growth. Therefore, it is recommended to use biochar as a soil amendment for long-term carbon sink restoration.

Author details

Jyoti Rawat[1], Jyoti Saxena[2*] and Pankaj Sanwal[2]

1 Department of Biotechnology, Kumaun University, Bhimtal Campus, Nainital, India

2 Biochemical Engineering Department, B.T. Kumaon Institute of Technology, Dwarahat, India

*Address all correspondence to: saxenajyoti30@gmail.com

References

[1] Thalmann M, Santelia D. Starch as a determinant of plant fitness under abiotic stress. The New Phytologist. 2017;**214**(3):943-951

[2] Moon DH, Park JW, Chang YY, Ok YS, Lee SS, Ahmad M. Immobilization of lead in contaminated firing range soil using biochar. Environmental Science and Pollution Research. 2013;**20**:8464-8471

[3] Al-Farraj AS, Usman ARA, Al Otaibi SHM. Assessment of heavy metals contamination in soils surrounding a gold mine: Comparison of two digestion methods. Chemistry and Ecology. 2013;**29**:329-339

[4] Noman A, Aqeel M. miRNA-based heavy metal homeostasis and plant growth. Environmental Science and Pollution Research. 2017;**24**(11):3-10

[5] Pathak H. Trend of fertility status of Indian soils. Current Advances in Agricultural Sciences. 2010;**2**(1):10-12

[6] Hariprasad NV, Dayananda HS. Environmental impact due to agricultural runoff containing heavy metals—A review. International Journal of Scientific and Research Publications. 2013;**3**(5):1-6

[7] Yargholi B, Azarneshan S. Long-term effects of pesticides and chemical fertilizers usage on some soil properties and accumulation of heavy metals in the soil (case study of Moghan plain's (Iran) irrigation and drainage network). International Journal of Agriculture and Crop Sciences. 2014;**7**:518-523

[8] Prashar P, Shah S. Impact of fertilizers and pesticides on soil microflora in agriculture sustainable agriculture reviews. Sustainable Agriculture Reviews. 2016;**19**:331-362

[9] Wu F, Gai Y, Jiao Z, Liu Y, Ma X, An L, et al. The community structure of microbial in arable soil under different long-term fertilization regimes in the Loess Plateau of China. African Journal of Microbiology Research. 2012;**6**:6152-6164

[10] Paz-Ferreiro J, Lu H, Fu S, Méndez A, Gascó G. Use of phytoremediation and biochar to remediate heavy metal polluted soils: A review. Solid Earth Discussions. 2013;**5**:2155-2179

[11] Lahori AH, Zhanyu G, Zhang Z, li R, Mahar A, Awasthi M, et al. Use of biochar as an amendment for remediation of heavy metal-contaminated soils: Prospects and challenges. Pedosphere. 2017;**27**:991-1014

[12] García AC, de Souza LGA, Pereira MG, Castro RN, García-Mina JM, Zonta E, et al. Structure-property-function relationship in humic substances to explain the biological activity in plants. Scientific Reports. 2016;**6**:20798. DOI: 10.1038/srep20798

[13] Zhang R, Zhang Y, Song L, Song X, Hanninen H, Wu J. Biochar enhances nut quality of *Torreya grandis* and soil fertility under simulated nitrogen deposition. Forest Ecology and Management. 2017;**391**:321-329

[14] Debela F, Thring RW, Arocena JM. Immobilization of heavy metals by co-pyrolysis of contaminated soil with woody biomass. Water, Air, and Soil Pollution. 2012;**223**:1161-1170

[15] Bonanomi G, Ippolito F, Cesarano G, Nanni B, Lombardi N, Rita A, et al. Biochar as plant growth promoter: Better off alone or mixed with organic amendments? Frontiers in Plant Science. 2017;**8**:1570. DOI: 10.3389/fpls.2017.01570

[16] Dume B, Mosissa T, Nebiyu A. Effect of biochar on soil properties

and lead (Pb) availability in a military camp in South West Ethiopia. African Journal of Environmental Science and Technology. 2016;**10**(3):77-85

[17] Verheijen F, Jeffery S, Bastos AC, van der Velde M, Diafas I. Biochar Application to Soils—A Critical Scientific Review of Effects on Soil Properties, Processes and Functions. Luxemburg: EUR 24099 EN Office for the Official Publications of the European Communities; 2009. p. 149

[18] Nigussie A, Kissi E, Misganaw M, Ambaw G. Effect of biochar application on soil properties and nutrient uptake of Lettuces (*Lactuca sativa*) grown in chromium polluted soils. American-Eurasian Journal of Agriculture and Environmental Science. 2012;**12**(3):369-376

[19] Chan K, Van Zwieten L, Meszaros I, Downie A, Joseph S. Using poultry litter biochars as soil amendments. Australian Journal of Soil Research. 2008;**46**(5):437-444

[20] Saxena J, Rana G, Pandey M. Impact of addition of biochar along with *Bacillus* sp. on growth and yield of French beans. Scientia Horticulturae. 2013;**162**:351-356

[21] Placek A, Grobelak A, Kacprzak M. Improving the phytoremediation of heavy metals contaminated soil by use of sewage sludge. International Journal of Phytoremediation. 2016;**18**(6):605-618

[22] Ronsse F, van Hecke S, Dickinson D, Prins W. Production and characterization of slow pyrolysis biochar: Influence of feedstock type and pyrolysis conditions. GCB Bioenergy. 2013;**5**:104-115

[23] Kiran YK, Barkat A, Xiao-qiang CUI, Ying F, Feng-shan P, Lin T, et al. Cow manure and cow manure-derived biochar application as a soil amendment for reducing cadmium availability and accumulation by *Brassica chinensis* L. in acidic red soil. Journal of Integrative Agriculture. 2017;**16**(3):725-734

[24] Chan KY, Van Zwieten L, Meszaros I, Downie A, Joseph S. Agronomic values of greenwaste biochar as a soil amendment. Australian Journal of Soil Research. 2007;**45**:629-634

[25] Novotny EH, Maia CMB de F, Carvalho MT de M, Madari BE. Biochar: Pyrogenic carbon for agricultural use—A critical review. Revista Brasileira de Ciência do Solo. 2015;**39**(2):321-344

[26] Kameyama K, Miyamoto T, Iwata Y, Shiono T. Influences of feedstock and pyrolysis temperature on the nitrate adsorption of biochar. Soil Science and Plant Nutrition. 2016;**62**(2):180-184

[27] Woolf D, Amonette J, Street-Perrott F, Lehmann J, Joseph S. Sustainable biochar to mitigate global climate change. Nature Communications. 2010;**1**:1-9

[28] Tumuluru JS, Sokhansanj S, Hess JR, Wright CT, Boardman RD. A review on biomass torrefaction process and product properties for energy applications. Industrial Biotechnology. 2011;**7**:384-402

[29] Sohi S, Loez-Capel S, Krull E, Bol R. Biochar's roles in soil and climate change: A review of research needs. CSIRO Land and Water Science Report. 2009;**5**(09):17-31

[30] Reddy KR. Characteristics and applications of biochar for environmental remediation: A review. Critical Reviews in Environmental Science and Technology. 2015;**45**:939-969

[31] Peterson SC, Jackson MA, Kim S, Palmquist DE. Increasing biochar surface area: Optimization of ball milling parameters. Powder Technology. 2012;**228**:115-120

[32] Sullivan AL, Ball R. Thermal decomposition and combustion chemistry of cellulosic biomass. Atmospheric Environment. 2012;**47**:133-141

[33] Zhu L, Lei H, Zhang Y, Zhang X, Bu Q, Wei Y, et al. A review of biochar derived from pyrolysis and its application in biofuel production. SF Journal of Material and Chemical Engineering. 2018;**1**(1):1007

[34] Brown TR, Wright MM, Brown RC. Estimating profitability of two biochar production scenarios: Slow pyrolysis vs fast pyrolysis. Biofuels, Bioproducts and Biorefining. 2011;**5**(1):54-68

[35] Joseph SD, Camps-Arbestain M, Lin Y, Munroe P, Chia CH, Hook J, et al. An investigation into the reactions of biochar in soil. Australian Journal of Soil Research. 2010;**48**:501-515

[36] Antal MJ, Grønli M. The art, science, and technology of charcoal production. Industrial and Engineering Chemistry Research. 2003;**42**(8):1619-1640

[37] Peng X, Ye LL, Wang CH, Bo S. Temperature and duration dependent rice straw derived biochar: Characteristics and its effects on soil properties of an Ultisol in southern China. Soil and Tillage Research. 2011;**112**(2):159-166

[38] Dutta B. Assessment of pyrolysis techniques of lignocellulosic biomass for biochar production [dissertation—master's thesis]. McGill University; 2010

[39] Stoyle A. Biochar production for carbon sequestration [master's thesis]. Shanghai Jiao Tong University; 2011

[40] Shenbagavalli S, Mahimairaja S. Production and characterization of biochar from different biological wastes. International Journal of Plant, Animal and Environmental Sciences. 2012;**2**(1):197-201

[41] Ajema L. Effects of biochar application on beneficial soil organism review. International Journal of Research Studies in Science, Engineering and Technology. 2018;**5**(5):9-18

[42] Liang B, Lehmann J, Solomon D, Kinyangi J, Grossman J, O"Neill B, et al. Black carbon increases cation exchange capacity in soils. Soil Science Society of America Journal. 2006;**70**:1719-1730

[43] Lehmann J. Bio-energy in the black. Frontiers in Ecology and the Environment. 2007;**5**(7):381-387

[44] Blanco-Canqui H. Biochar and soil physical properties. Soil Science Society of America Journal. 2017;**81**:687-711

[45] Hammes K, Schmidt MWI. Changes in biochar in soils. In: Lehmann M, Joseph S, editors. Biochar for Environmental Management Science and Technology. London: Earthscan; 2009. pp. 169-182

[46] Lehmann J, Gaunt J, Rondon M. Biochar sequestration in the terrestrial ecosystem—A review. Mitigation and Adaptation Strategies for Global Change. 2006;**11**:403-427

[47] DeLuca TH, Gundale MJ, MacKenzie MD, Jones DL. Biochar effects on soil nutrient transformation. In: Lehmann J, Joseph S, editors. Biochar for Environmental Management: Science and Technology. 2nd ed. Routledge; 2015. pp. 421-454

[48] Cao Y, Gao Y, Qi Y, Li J. Biochar-enhanced composts reduce the potential leaching of nutrients and heavy metals and suppress plant-parasitic nematodes in excessively fertilized cucumber soils. Environmental Science and Pollution Research International. 2018;**25**(8):7589-7599

[49] Slapakova B, Jerabkova V, Tejnecky DO. The biochar effect on soil respiration and nitrification. Plant, Soil and Environment. 2018;**64**(3):114-119

[50] Saito M, Marumoto T. Inoculation with arbuscular mycorrhizal fungi: The status quo in Japan and the future prospects. Plant and Soil. 2002;**244**:273-279

[51] Skjemstad JO, Janik LJ, Taylor JA. Non-living soil organic matter: What do we know about it? Australian Journal of Experimental Agriculture. 1998;**38**:667-680

[52] Swift RS. Sequestration of carbon by soil. Soil Science. 2001;**166**:858-871

[53] Krull ES, Skjemstad J, Graetz D, Grice K, Dunning W, Cook G, et al. 13C-depleted charcoal from C4 grasses and the role of occluded carbon in phytoliths. Organic Geochemistry. 2003;**34**:1337-1352

[54] Hazarika BN, Ansari S. Biofertilizers in fruit crops—A review. Agricultural Reviews. 2007;**28**(1):69-74

[55] Beesley L, Moreno-Jiménez E, Gomez-Eyles JL, Harris E, Robinson B, Sizmur T. A review of biochars' potential role in the remediation, revegetation and restoration of contaminated soils. Environmental Pollution. 2011;**159**:3269-3282

[56] Mensah AK, Frimpong KA. Biochar and/or compost applications improve soil properties, growth, and yield of maize grown in acidic rainforest and coastal savannah soils in Ghana. International Journal of Agronomy. 2018:1-8. DOI: 10.1155/2018/6837404

[57] Glaser B, Lehmann J, Zech W. Ameliorating physical and chemical properties of highly weathered soils in the tropics with charcoal: A review. Biology and Fertility of Soils. 2002;**35**:219-230

[58] Lehmann J, da Silva JP Jr, Steiner C, Nehls T, Zech W, Glaser B. Nutrient availability and leaching in an archaeological anthrosol and a ferrasol of the Central Amazon basin: Fertilizer, manure, and charcoal amendments. Plant and Soil. 2003;**249**:343-357

[59] Yuan J, Xu R, Zhang H. The forms of alkalis in the biochar produced from crop residues at different temperatures. Bioresource Technology. 2011;**102**:3488-3497

[60] Zhang A, Cui L, Pan G, Li L, Hussain Q, Zhang X, et al. Effect of biochar amendment on yield and methane and nitrous oxide emissions from a rice paddy from Tai Lake plain, China. Agriculture, Ecosystems and Environment. 2010;**139**(4):469-475

[61] Spokas KA, Koskinen WC, Baker JM, Reicosky DC. Impacts of woodchip biochar additions on greenhouse gas production and sorption/degradation of two herbicides in a minnesota soil. Chemosphere. 2009;**77**:574-581

[62] Bonanomi G, Ippolito F, Scala F. A "black" future for plant pathology? Biochar As a new soil amendment for controlling plant diseases. Journal of Plant Pathology. 2015;**97**(2):223-234

[63] Elmer W, White JC, Pignatello JJ. Impact of biochar addition to soil on the bioavailability of chemicals important in agriculture. Report. New Haven: University of Connecticut; 2010

[64] Thies JE, Rillig M. Characteristics of biochar: Biological properties. In: Lehmann M, Joseph S, editors. Biochar for Environmental Management Science and Technology. London: Earthscan; 2009. pp. 85-105

[65] Nerome M, Toyota K, Islam TM, Nishimima T, Matsuoka T, Sato K, et al. Suppression of bacterial wilt of tomato by incorporation of municipal biowaste

charcoal into soil. Soil Microorganisms. 2005;**59**(1):9-14

[66] Ogawa M. Charcoal use in agriculture in Japan. Keynote address, 1st Asia Pacific Biochar Conference, May 17-20, 2009, Gold Coast, Australia

[67] van Zwieten L, Kimber S, Downie A, Morris S, Petty S, Rust J, et al. A glasshouse study on the interaction of low mineral ash biochar with nitrogen in a sandy soil. Australian Journal of Soil Research. 2010;**48**:569-576

[68] Yu XY, Ying GG, Kookana RS. Reduced plant uptake of pesticides with biochar additions to soil. Chemosphere. 2009;**76**:665-671

[69] Hua L, Wu WX, Liu YX, McBride M, Chen YX. Reduction of nitrogen loss and Cu and Zn mobility during sludge composting with bamboo charcoal amendment. Environmental Science and Pollution Research International. 2009;**16**:1-9

[70] Major J, Lehmann J, Rondon M, Goodale C. Fate of soil-applied black carbon: Downward migration, leaching and soil respiration. Global Change Biology. 2010;**16**(4):1366-1379

[71] Singh PB, Hatton JB, Singh B, Cowie LA, Kathuria A. Influence of biochars on nitrous oxide emission and nitrogen leaching from two contrasting soils. Journal of Environmental Quality. 2010;**39**:1224-1235

[72] Xie Z, Xu Y, Liu G, Liu Q, Zhu J, Tu C, et al. Impact of biochar application on nitrogen nutrition of rice, greenhouse-gas emissions and soil organic carbon dynamics in two paddy soils of China. Plant and Soil. 2013;**370**(1-2):527-540

[73] Taghizadeh-Toosi A, Clough TJ, Sherlock RR, Condron LM. Biochar adsorbed ammonia is bioavailable. Plant and Soil. 2012;**350**:57-69

[74] Saarnio S, Heimonen K, Kettunen R. Biochar addition indirectly affects N_2O emissions via soil moisture and plant N uptake. Soil Biology and Biochemistry. 2013;**58**:99-106

[75] Rondon M, Lehmann J, Ramírez J, Hurtado M. Biological nitrogen fixation by common beans (*Phaseolus vulgaris* L.) increases with bio-char additions. Biology and Fertility of Soils. 2007;**43**:699-708

[76] Gaskin JW, Speir RA, Harris K, Das KC, Lee RD, Morris LA, et al. Effect of peanut hull and pine chip biochar on soil nutrients, corn nutrient status, and yield. Agronomy Journal. 2010;**102**:623-633

[77] Nartey OD, Zhao B. Biochar preparation, characterization, and adsorptive capacity and its effect on bioavailability of contaminants: An overview. Advances in Materials Science and Engineering. 2014;**2014**(715398). p. 12. Available from: http://dx.doi.org/10.1155/2014/715398

[78] Busscher WJ, Novak JM, Evans DE, Watts DW, Niandou MAS, Ahmedna M. Influence of pecan biochar on physical properties of a Norfolk loamy sand. Soil Science. 2010;**175**:10-14

[79] Saxena J, Rawat J, Sanwal P. Enhancement of growth and yield of *glycine max* plants with inoculation of phosphate solubilizing fungus *Aspergillus niger* K7 and biochar amendment in soil. Communications in Soil Science and Plant Analysis. 2016;**47**(20):2334-2347

[80] Saxena J, Rawat J, Kumar R. Conversion of biomass waste into biochar and the effect on mung bean crop production. Clean—Soil, Air, Water. 2017;**45**(7):1501020 (1-9)

[81] Solaiman ZM, Blackwell P, Abbott LK, Storer P. Direct and residual effect of biochar application on mycorrhizal

root colonisation, growth and nutrition of wheat. Australian Journal of Soil Research. 2010;**48**:546-554

[82] Warnock DD, Lehmann J, Kuyper TW, Rillig MC. Mycorrhizal responses to biochar in soil—Concepts and mechanisms. Plant and Soil. 2007;**300**:9-20

[83] Rondon M, Ramirez A, Hurtado M. Charcoal Additions to High Fertility Ditches Enhance Yields and Qvuality of Cash Crops in Andean Hillsides of Columbia (CIAT Annual Report 2004). Cali, Colombia; 2004

[84] Schwartz MW, Hoeksema JD, Gehring CA, Johnson NC, Klironomos JN, Abbott LK, et al. The promise and the potential consequences of the global transport of mycorrhizal fungal inoculum. Ecology Letters. 2006;**9**:501-515

[85] Saito M. Charcoal as a micro habitat for VA mycorrhizal fungi, and its practical application. Agriculture, Ecosystems and Environment. 1990;**29**:341-344

[86] Pietikainen J, Kiikkila O, Fritze H. Charcoal as a habitat for microbes and its effect on the microbial community of the underlying humus. Oikos. 2000;**89**:231-242

[87] Ezawa T, Yamamoto K, Yoshida S. Enhancement of the effectiveness of indigenous arbuscular mycorrhizal fungi by inorganic soil amendments. Soil Science and Plant Nutrition. 2002;**48**:897-900

[88] Major J, Steiner C, Ditommaso A, Falcão NP, Lehmann J. Weed composition and cover after three years of soil fertility management in the central Brazilian Amazon: Compost, fertilizer, manure and charcoal applications. Weed Biology and Management. 2005;**5**:69-76

[89] Oguntunde PG, Fosu M, Ajayi AE, van de Giesen N. Effects of charcoal production on maize yield, chemical properties and texture of soil. Biology and Fertility of Soils. 2004;**39**:295-299

[90] Major J, Rondon M, Molina D, Riha SJ, Lehmann J. Maize yield and nutrition during four years after biochar application to a Colombian savanna oxisol. Plant and Soil. 2010;**333**:117-128

[91] Asai H, Samson BK, Stephan HM, Songyikhangsuthor K, Homma K, Kiyono Y, et al. Biochar amendment techniques for upland rice production in Northern Laos: 1. Soil physical properties, leaf SPAD and grain yield. Field Crops Research. 2009;**111**(1-2):81-84

[92] Silber A, Levkovitch I, Graber ER. pH-dependent mineral release and surface properties of cornstraw biochar: Agronomic implications. Environmental Science and Technology. 2010;**44**:9318-9323

[93] Mikan CJ, Abrams MD. Altered forest composition and soil properties of historic charcoal hearths in southeastern Pennsylvania. Canadian Journal of Forest Research. 1995;**25**:687-696

[94] Matsubara Y, Hasegawa N, Fukui H. Incidence of *Fusarium* root rot in asparagus seedlings infected with arbuscular mycorrhizal fungus as affected by several soil amendments. Journal of the Japanese Society For Horticultural Science. 2002;**71**:370-374

[95] Elmer WH, Pignatello JJ. Effect of biochar amendments on mycorrhizal associations and *Fusarium* crown and root rot of asparagus in replant soils. Plant Disease. 2011;**95**:960-966

[96] Novak JM, Lima I, Gaskin JW, Steiner C, Das KC, Ahmedna M, et al. Characterization of designer biochar produced at different temperatures and their effects on a loamy sand. Annals of Environmental Science. 2009;**3**:195-206

[97] Kammann C, Glaser B, Schmidt HP. Combining biochar and organic amendments. In: Shackley S, Ruysschaert G, Zwart K, Glaser B, editors. Biochar in European Soils and Agriculture: Science and Practice. New York: Routledge; 2016. pp. 136-164

[98] Liu J, Schulz H, Brandl S, Miehtke H, Huwe B, Glaser B. Short-term effect of biochar and compost on soil fertility and water status of a Dystric Cambisol in NE Germany under field conditions. Journal of Plant Nutrition and Soil Science. 2012;**175**(5):1-10

[99] Yanai Y, Toyota K, Okazaki M. Effects of charcoal addition on N_2O emissions from soil resulting from rewetting air-dried soil in short-term laboratory experiments. Soil Science and Plant Nutrition. 2007;**53**:181-188

[100] Spokas KA, Reikosky DC. Impacts of sixteen different biochars on soil greenhouse gas production. Annals of Environmental Science. 2009;**3**:179-193

Chapter 2

Biochar and Soil Physical Health

Kayode S. Are

Abstract

The use of organic materials for reclamation of soil physical health indicators of degraded soil is germane for sustainable agriculture. Despite the soil conservation effectiveness of organic fertilizer, its adoption remains low among smallholder farmers in most parts of sub-Saharan Africa because of its offensive odor and bulkiness. Farmers desire materials that are not bulky, handled with ease, ensure maximum nutrient retention, improve soil structural quality, reduce soil compaction, and increase water retention, which will also increase soil productivity and crop yield. These are the greatest attractions for the introduction of biochar for improvement of soil physical health. The pyrolytic processes of various organic materials to biochar have suppressed the effects of distractive odor of fresh and composted organic materials while reducing the bulkiness experienced during application. The potentials of biochar in improving nutrient retention and release have been published by various authors, but little information is available for soil physical health indicators. Therefore, the potentials of biochar in restoring physical health indicators such as particle size distribution, bulk density, pore size distribution, soil water retention and distribution, compaction and aggregate size distribution and stability of degraded soil shall be discussed in this chapter.

Keywords: degraded soil, biochar, physical health indicator, soil productivity, sustainable agriculture

1. Introduction

Soil physical health is the ability of a given soil to meet plant and ecosystem requirements for water, aeration, and strength over time and to resist and recover from processes that might diminish that ability [1]. Application of organic materials for soil amendment, especially the composted manures, plays important roles in reclaiming and improving the physical health of degraded soils [2]. They have profound influence on almost all soil properties—such as structure (and hence on water infiltration and storage, susceptibility to surface runoff and erosion), cation exchange capacity, nutrient availability, buffering (pH, nutrient availability), color, and plant pest pressure. In spite of these potentials, their adoption as soil amendment remains low among smallholder farmers in most parts of sub-Saharan Africa because of their offensive odor and bulkiness. However, one of the greatest attractions for the use of biochar is the suppression of the effects of distractive odor of fresh and composted organic materials through pyrolytic processes, while the bulkiness experienced during application of composted manure is reduced.

Biochar is a carbon-rich organic matter, which is generally derived from the incomplete combustion of waste biomass, and it is produced by the slow thermochemical pyrolysis of biomass materials. Organic wastes, such as livestock manures, sewage sludge, crop residues, and composts are converted to biochars and then applied to soils as an amendment. Biochar application as soil amendment improves crop productivity, enhances soil properties, and increases carbon storage in the soil due to its highly recalcitrant carbon content [3]. This practice has, however, received a growing interest as a sustainable process to improve the properties of highly degraded tropical soils [4, 5]. Biochars are characteristically very light materials with a high porosity and surface area, which alter some soil physical properties such as the bulk density (BD), water-holding capacity (WHC), surface area, and penetration resistance (PR) [6]. In Nigeria, when comparing the potential of poultry biochar with composted and noncomposted poultry litter, Are et al. [2] recorded an increase in soil water retention of between 3.3 and 31.3% following application of poultry litter biochar than uncharred poultry manures at lower water suction. Elsewhere, Major et al. [7] reported that the surface soils of oxisols amended with char at 20 Mg ha^{-1} contained more water by volume, and the water was held more tightly than unamended soils. In China, Chen [8] reported a decrease in bulk density by 4.5 and 6% with addition of 2.25 and 4.50 Mg ha^{-1}, respectively, while an increase in water holding capacity from 25 to 36% was recorded by Kinney [9] with 7% biochar by weight addition.

In spite of the benefits of biochar on soil physical properties reported by different authors [2, 6–9], most positive effects of biochar are seen with coarse- or medium-textured soils, suggesting improvement of water holding capacity (WHC) by biochar addition [10] but not with fine-textured soils. Research has shown that unfavorable soil physical changes sometimes occur when biochar is added as soil amendment. Soil aggregation, for instance, may not be immediately enhanced by biochar addition [6]. The application of oak-650 biochar (0.5%, w/w) by Mukherjee and Lal [11] on a degraded silty clay loam soil reduced aggregation by 10% relative to the control. Mukherjee [11] suggested that (i) there may be a threshold application rate below which no aggregate stability is achieved, and/or (ii) a higher interaction time is required. On the other hand, Tryon [12] reported that application of pine (Pinus spp.) and oak (Quercus spp.) biochars increased available water content (AWC) in a sandy soil, while having no effect in a loamy soil, and it decreased moisture content in a clayey soil, indicating that the effect of biochar on AWC can be strongly influenced by the soil textural classes. Similarly, Masiello et al. [13] reported that a high rate (up to 11.3 Mg ha^{-1}) of maize stover biochar pyrolyzed at 350 and 550°C did not improve AWC in amended silt loam soils after incubation for 295 days, which was attributed to clogging of micropores by ash over time. The contrasting behaviors of biochars have been attributed by various researchers to biochar's particle size, shape, and internal structure, which alter pore characteristics and consequently influence soil water storage. With these contrasting trends (both positive and negative) of future biochar, future studies, especially at field scale with similar soil types with different biochar combinations over time, may shed light on this aspect. This chapter will discuss the practical use of biochar as it relates to the overall soil physical health.

2. Physical properties of biochar

Biochar is difficult to classify based on its properties, both chemical and physical, because of the variability imparted to it by the production conditions (time, temperature) and feedstock. Biochars (**Figure 1**) are of different particle sizes and do not have the same properties since their characteristics are controlled by many factors. Operating factors during the pyrolysis process that influence the resultant physical properties of biochar of any given biomass feedstock include heating rate, highest treatment temperature, pressure, reaction residence time, reaction vessel (orientation,

Figure 1.
Biochars from feedstocks with different particle sizes.

dimensions, stirring regime, catalysts, etc.), pretreatment (drying, comminution, chemical activation, etc.), the flow rate of ancillary inputs (e.g., nitrogen, carbon dioxide, air, steam, etc.), and posttreatment (crushing, sieving, activation, etc.).

Although all of these parameters contribute to the final biochar structure, the pyrolysis highest treatment temperature has been identified by Downie et al. [14] as the most important of the factors since the fundamental physical changes (i.e., the release of volatiles, the formation of intermediate melts, and the volatilization of the intermediate melts) are all temperature dependent. The temperature ranges, however, under which these stages occur, vary with feedstock. Heating rates and pressures are expected to have the second greatest influence since they affect the physical mass transfer of volatiles evolving at the given temperature from the reacting particles [15, 16].

An additional mechanism producing the structural complexity of biochars is the occurrence of cracking. Biochar is typically laced with macrocracks, which can be related to both feedstock properties and the rate at which carbonization is carried out [17]. Wood biochar is generally broken and cracked due to shrinkage stresses developed because the surface of the material decomposes faster than its interior. Brown et al. [18] concluded that high-temperature (1000°C) surface area is controlled primarily by low-temperature (<450°C) cracking and high-temperature microstructural rearrangement.

The physical characteristics can be both directly and indirectly related to the way in which they affect soil systems. The physical characteristics of biochar depend not only upon the starting organic material (biomass), but also upon the carbonization or pyrolysis system by which they are made (including the pre- and posthandling of the biomass and biochar) [14].

The fundamental molecular structure of biochar creates both its surface area and porosity. However, pyrolysis processing of biomass enlarges the crystallites and makes them more ordered. This effect increases with highest treatment temperature. Lua et al. [15] demonstrated that increasing the pyrolysis temperature from 250 to 500°C increases the Brunauer, Emmett, and Teller equation (BET) surface area due to the increasing evolution of volatiles from pistachio-nut shells, resulting in enhanced pore development in biochars. For turbostratic arrangements, the successive layer planes are disposed approximately parallel and equidistant, but rotated more or less randomly with respect to each other (**Figure 2**). The spacing between the planes of turbostratic regions of biochar is larger than that observed in graphite [19].

In relating biochars with soil physical properties, biochar's particle size, shape, and internal structure play important roles in controlling soil water storage because they alter pore characteristics. For instance, biochar has pores inside particles (intrapores), which may provide additional space for water storage beyond the pore space between particles (interpores) [20]. Particle size may influence both intrapores and interpores through different processes because the size and connectivity of these particles likely differ. In addition, when applied in the field, biochar particles may have different sizes and shapes compared to soil particles. This addition of biochar grains with different shapes and sizes will change interpore characteristics (size, shape, connectivity, and volume) of soil and thus will affect water storage and mobility. For instance, fine biochar particles can fill pores between coarse soil particles, decreasing pore size and changing interpore shape.

An important physical property of biochar is its stability in the environment. However, degradation of at least some components (such as volatile matter or labile OM) of biochar may occur [21, 22]. On the other hand, subsoils are characteristically different due to variations in microbial activity and oxygen content, which affect biochar oxidation and aging.

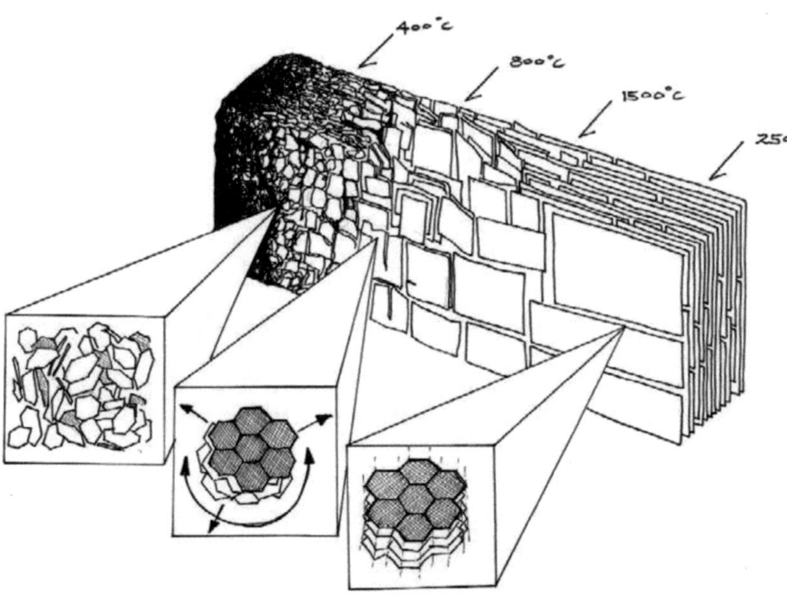

Figure 2.
Ideal biochar structure development with highest treatment temperature (HTT): (a) increased proportion of aromatic C, highly disordered in amorphous mass; (b) growing sheets of conjugated aromatic carbon, turbostratically arranged; and (c) structure becomes graphitic with order in the third dimension (source: [14]).

3. Soil physical health and biochars

3.1 What is soil physical health?

Soil health synonymous to soil quality is usually considered to have three main aspects: physical, chemical, and biological. It is considered to be important for the assessment of the extent of land degradation or amelioration, and for identifying management practices for sustainable land use. However, the knowledge of the physical properties of soil is essential for improving soil health to achieve optimal productivity for each soil type in a given climatic condition. According to Dexter [23], soil physical health manifested in various ways. For instance, soils poor physical health are those that exhibit one or more of the following symptoms: poor water infiltration, run-off of water from the surface, hard-setting, poor aeration, poor rootability, and poor workability. On the other hand, good soil physical health occurs when soils exhibit the opposite or the absence of the conditions listed above. However, there has been no single measure of soil physical health [24] but an integration of a range of some physical properties to obtain an overall assessment.

3.2 Soil physical health indicators as affected by biochar amendment

As mentioned earlier in this chapter, the effects of biochar on soil physical health indicators depend on several factors, such as biomass or feedstock type, pyrolytic condition, application rate, and environmental condition. The effects of biochar-amended soil in relation to some physical properties are discussed here.

3.2.1 Soil surface area

Surface area is an important soil physical health indicator that influences essential functions of soil fertility, including water, nutrient retention, aeration, and microbial activity [14]. For instance, the limited capacity of sandy soil to store water and plant nutrients is partly related to the relatively small surface area of its soil particles [25]. Coarse sands have a very low specific surface area of about 0.01 $m^2 g^{-1}$, compared to fine sands of 0.1 $m^2 g^{-1}$ and clays' large specific surface area ranging from 5 $m^2 g^{-1}$ for kaolinite to about 750 $m^2 g^{-1}$ for Na-exchanged montmorillonite [25]. Therefore, soils containing a large fraction of clay may have high total water-holding capacities but inadequate aeration. Meanwhile, Troeh and Thompson [25] reported that high organic matter contents have the potentials to overcome the problem of too much water held in a clay soil, while increasing the water contents in a sandy soil. However, studies have shown that biochar will similarly change the physical nature of soil, having much of the same benefit of other organic amendments in this regard [2, 26]. Biochar-specific surfaces, being generally higher than sand and comparable to or higher than clay, will therefore cause a net increase in the total soil-specific surface when added as an amendment [14]. The high surface area of biochar provides space for formation of bonds and complexes with cations and anions with metals and elements of soil on its surface, which may improve the water and nutrient retention capacity of soil. A long-term soil column incubation study by Laird et al. [27] indicated increases in specific surface area of an amended clayey soil from 130 to 153 $m^2 g^{-1}$ as the biochar concentration increased from 0 to 20 $g\ kg^{-1}$.

3.2.2 Bulk density and pore-size distribution

Many studies have observed decreases in bulk density and increases in porosity as a result of biochar application [2, 6, 7, 26, 28]. Roughly, 2% (by weight) of biochar in soil is an enough addition to show a significant decrease in bulk density in amended soils [6, 7]. The rate of biochar application as well as the density and porosity of the original soil are critical in predicting the effects of biochar addition to any soil. Using peanut hulls, Githinji [28] recorded reductions in bulk density with increased rate of biochar amendment, and he [28] recorded the highest bulk density of 1.33 g cm^{-3} for the soil without biochar amendment, decreasing to 1.09 g cm^{-3} for 25% rate, 0.89 g cm^{-3} for 50% rate, 0.61 g cm^{-3} for 75% rate, and 0.36 g cm^{-3} for 100% rate of biochar application. Since bulk density is a measure of the relative mass of a solid relative to the bulk volume the solid occupies, including the void spaces, it follows that the greater is the portion occupied by the pores, the lower is the bulk density of a solid. The upper limit of the bulk density would be a situation where there are no pores, and this limit will approach that of particle density of a solid.

The relationship between total surface area and pore-size distribution is logical. It is logical that this physical feature of biochars will also be of importance to their behavior in soil processes. As shown in **Figure 2**, the increase in HTT results in more structured regular spacing between the planes. Interplanar distances also decrease with the increased ordering and organization of molecules, all of which result in larger surface areas per volume. Githinji [28] reported that for the nonamended soil, porosity was 0.50 cm^3 cm^{-3}, increasing to 0.55, 0.61, 0.69, and 0.78 cm^3 cm^{-3}, respectively, for 25, 50, 75, and 100% rates of biochar application. In another trial comparing poultry litter biochar-amended soil and uncharred poultry manure, Are et al. [2] recorded a significance increase in storage pores (0.5–50 μm equivalent cylindrical radius) of a biochar-amended soil than uncharred poultry manure. However, this was not the case of transmission pores, where the soil amended with poultry biochar had lower transmission pores than uncharred poultry materials [2]. Mesoporosity may also increase significantly at the expense of macropores in waste-derived biochar-amended soil compared to control, with the higher rate of biochar application having a greater effect [29].

3.2.3 Soil water retention

The quantification of the amount of water held at field capacity (θfc) and at permanent wilting point (θpwp), and the amount of plant available water (θpaw) of soil with biochar amendment is an efficient way to quantify how biochar affects soil water conditions and plant growth. Previous studies have shown that biochar increased water retention of soil [7, 30]. Gaskin et al. [31] reported a doubling in the mean volumetric water content of a loamy sandy soil at 2 kPa following the application of peanut hull biochar at a rate of 88 t/ha. Whereas Are et al. [2] also reported as high as 33% change in moisture content with application of poultry litter biochar to a sandy loam soil. However, the mechanisms controlling these observations should be understood. Sandy soils, which have larger pore space, are particularly appealing target for biochar amendment because studies on sand and sandy loam often show an increase in plant available water after biochar amendment [32, 33]. However, few studies focused on the mechanism of how biochar increase the available water. Without understanding the mechanisms that control biochar-driven changes of water retention of soil, it is difficult to predict when and by how much biochar will improve soil water retention.

Biochar's particle size, shape, and internal structure may play important roles in controlling soil water storage because they alter pore characteristics. For instance, biochar has pores inside particles (intrapores), which may provide additional space for water storage beyond the pore space between particles (interpores) [20]. Particle size may influence both intrapores and interpores through different processes because the size and connectivity of these particles likely differ. Intraporosity increases plant available water, suggesting that biochar with high intraporosity will be most useful. Feedstock type, pyrolysis temperature, and charring residence time influence biochar's intraporosity [34]. Biochars with low intraporosity such as wastewater sludge biochar and poultry litter biochar are less favorable for soil water storage at low water potentials (<−16.5 kPa) because their internal porosity is very low [35]. In addition, the efficiency of biochar for improving soil water retention will be reduced if biochars are hydrophobic, but hydrophobicity can likely be managed by pretreatment [21]. Hydrophobic biochar has positive water entry pressure, meaning that an applied force is required for water to enter intrapores. Biochar hydrophobicity can prevent water from penetrating into biochar intrapores, prohibiting an improvement of soil water retention [10]. This indicates that biochars with low hydrophobicity will enhance soil water retention than those with high hydrophobic. Jeffery et al. [10] reported that grass species biochar did not improve soil water retention; this is probably due to its high hydrophobicity, although it is notable that grass biochar has lower hydrophobicity compared to leaf or wood biochars [9]. Biochar's hydrophobicity varies with production temperature and feedstock [36], but it is usually eliminated by brief environmental exposure. Pretreating biochar either by initially wetting it, or by composting is likely to significantly reduce problems associated with hydrophobicity [35].

3.2.4 Hydraulic conductivity

Hydraulic conductivity (K) measures the ease with which water can move through a soil, subject to a hydraulic gradient and is essential in infiltration-related applications such as irrigation and drainage management [37]. Saturated hydraulic conductivity (K_{sat}) is the conductivity measured, while the soil is saturated. In a trial in Ibadan, Nigeria, Are et al. [2] recorded a significant reduction in K_{sat} (9.2 mm h^{-1}) than other amendments (16.5–18.2 mm h^{-1}) in their poultry biochar trial. The reduction in the k_{sat} of poultry's biochar treatment soils was linked to the ash deposited by the biochar, which perhaps reduced the larger soil pores and thus led to the reduction in pore space and volumes. Several studies [2, 28, 38–40] have linked the reduction in soil hydraulic conductivity, especially sandy soil, to a reduction in porosity imposed by the fine-grained particles of biochar. Devaraux et al. [38] was of the opinion that the decrease was due to biochar's large surface area and the high number of pores, which had to be filled up before water drained under the force of gravity, meaning that more biochar in the soil might lead to the retention of more water in the storage pores. Barnes et al. [39], on the other hand, related shifts in K_{sat} to the physical mechanisms of the biochar, such as swelling and grain segregation, leading to the clogging of pores, decrease in pore radii, and possibly a variation in the bulk density and sample heterogeneity in the course of their experiment.

Contrasting results have been reported on the K_{sat} of a clay loam soil in Laos, following the application of biochar [40]. Asai et al. [40] reported a significant increase in K_{sat} on a clay loam soil with biochar amendment, whereas Major et al. [41] reported no significant effect in a clay soil following the addition of 20 t ha^{-1} biochar produced from wood. In a study by Barnes et al. [39], K_{sat} significantly increased in clay soil, decreased in sandy soil, and had no significant effect for sandy loam rich in organic matter following incorporation of biochar. The mixed results demonstrate that the interactions between applied biochar and soil amended with biochar, and the resulting effects on hydraulic conductivity are dependent on soil texture.

3.2.5 Soil aggregate stability and penetration resistance

Few data are available on aggregate stability and penetration resistance (PR) of biochar-amended soil. However, available information that exists is conflicting.

Examples of the few studies, which investigated soil aggregation with biochar amendment, are shown in **Table 1**. In a study by George et al. [42], the low-temperature (220°C) hydrochar made from spent brewer's grains, a residue from beer brewing, responded positively on aggregation of Albic Luvisol when (i) incubated for 5 months at 20°C in dark and (ii) used in a pot study with same hydrochar/soil combination (**Table 1**). These incubation and greenhouse studies involving plant indicate that hydrochar significantly increased water stable aggregates (WSA) compared to control, but the extent of WSA differed because the greenhouse study had 2–5 times higher rate of WSA formation compared to laboratory incubation. These data suggest that plant roots and mycorrhizal fungi, which were absent in the incubation study, had an important role in soil aggregation. In a field experiment, Are et al. [2] found that the poultry biochar amendment increased the WSA of a sandy loam soil from 41.6 to 59.1% of a four-season trial. In contrast, with and without mixing Bt and E horizons with pecan shell (*Carya illinoinensis*), biochar amendment decreased aggregation (**Table 1**) compared to control [43]. Mixing of biochar from pecan with switchgrass increased aggregation; however, the effect was significantly lower when soil was treated only with biochar and without mixing with switchgrass [44]. This trend indicates that a positive effect on soil aggregate stability requires presence of a substrate (i.e., switchgrass) along with biochar as an amendment. However, the application of biochar at the rate of 1% to an ultisol had no effect on aggregate stability [45]. Clearly, there exists limited information

Soil type	Biochar type	Study type (scale)	Rate of biochar application % ($g\ g^{-1}$)	Aggregation (%)		Penetration resistance (MPa)		Source
Norfolk loamy sand: E	Pecan shells, 700°C	Laboratory	0	14.3		1.19[a]	0.80[b]	[43]
			2.1	12.9		1.27[a]	0.88[b]	
Norfolk loamy sand: E and Bt			0	27.3		0.71[a]	0.76[b]	
			2.1	20.9		0.88[a]	0.94[b]	
Norfolk loamy sand: Ap	Pecan shells, 700°C	Laboratory	0	9.95	13.0*	1.04[a]	1.10[b]	[44]
			0.5	9.53	12.7*	0.96[a]	1.15[b]	
			1.0	10.7	12.3*	1.03[a]	1.02[b]	
			2.0	9.23	11.8*	0.82[a]	0.87[b]	
Albic Luvisol	Hydrochar, 220°C	Laboratory	0	49.8	—	—	—	[42]
			5	69.0	—	—	—	
			10	65.1	—	—	—	
		Greenhouse	0	10.3	—	—	—	
			5	20.8	—	—	—	
			10	33.8	—	—	—	
Alfisol		Field	0	41.6	—	—	—	[2]
			0.25	59.1	—	—	—	

[a]*Measured after 44 days.*
[b]*Measured after 96 days.*
**With switchgrass addition.*

Table 1.
Impact of biochar on aggregation and penetration resistance.

about how biochar affects aggregation and whether another substrate, plant roots, mycorrhizal fungi, or active-C source might be needed to increase WSA in biochar-amended soils. Nevertheless, the highest concentration of black-C was observed in the finest size fraction (<0.53 μm) of soil aggregates [46] suggesting preferential embedding of black-C particles compared to other organic compounds within aggregates. However, it was suggested by Jeffery et al. [10] that the hydrophobicity of biochar [10] may have increased the resistance of aggregates to slaking in water, which ultimately increased the aggregate stability.

The resistance of the soil to root penetration as determined by cone penetration resistance (PR) may not be alleviated by biochar addition over short time period but may be altered in the long run as aging of biochar changes its properties [47]. Along with time, soil type is also an important factor because another study reported reduction in PR with application of the same biochar on a different soil type (Norfolk loamy sand Ap) [44]. Nevertheless, the effect of biochar amendment on soil aggregation and PR requires additional research by including variations in biochar and soil type.

4. Conclusions

This review synthesizes available data on soil physical health indicators as influenced by application of biochars. The physical properties of biochar products affect many of the functional roles that they may play in improving soil physical health and environmental management. The large variation of physical characteristics observed in different biochar products means that some will be more effective than others in certain applications. It is important that the physical characterization of biochars is undertaken before they are experimentally applied to environmental systems, and variations in outcomes may be correlated with these features. The pyrolysis temperature, charring time of biochar and most importantly, the particle size of biochar play important factors in order to implement any biochar amendment project. The higher the biochar pyrolysis temperature, the finer the particle size, and the higher are the bulk density and water retention. The relationship may be inverse in relation to soil hydraulic conductivity and pore size distribution. This, however, depends on the soil type. Evidence has shown that biochar with finer particles when applied to sandy soil will reduce the macropores and hydraulic conductivity, whereas, in a clayey soil, biochar with finer particles will increase the interpores and soil hydraulic conductivity. Application rates of 0.25–2% ($g\ g^{-1}$) biochar can significantly improve soil physical health in terms of water-stable aggregates and water retention.

Acknowledgements

I wish to acknowledge the support enjoyed from the Institute of Agricultural Research and Training, Ibadan, Nigeria, for providing enabling environment to carry out biochar research. My appreciation goes to Biochar Initiatives in Nigeria, and to Ms. Romina Skomersic, whose encouragement geared me up in submitting this chapter.

Conflict of interest

The author declares that there is no conflict of interest.

Author details

Kayode S. Are
Institute of Agricultural Research and Training, Obafemi Awolowo University, Moor Plantation, Ibadan, Nigeria

*Address all correspondence to: kayodeare@gmail.com

References

[1] McKenzie BM, Tisdall JM, Vance WH. Soil physical quality. In: Gliński J, Horabik J, Lipiec J, editors. Encyclopedia of Agrophysics. The Netherlands: Springer Science + Business Media B.V; 2011. pp. 770-777

[2] Are KS, Adelana AO, Fademi IO, Aina OA. Improving physical properties of degraded soil: Potential of poultry manure and biochar. Agriculture and Natural Resources. 2017;**51**:454-462

[3] Lehmann J, Joseph S. An introduction. In: Lehmann J, Joseph S, editors. Biochar for Environmental Management. London: Earthscan; 2009. pp. 1-9

[4] Lehmann J, Rondon M. Bio-char soil management on highly weathered soils in the humid tropics. In: Uphoff N, Ball AS, Palm C, Fernandes E, Pretty J, Herrren H, Sanchez P, Husson O, Sanginga N, Laing M, Thies J, editors. Biological Approaches to Sustainable Soil Systems. Boca Raton, FL: CRC Press; 2006. pp. 517-530

[5] Oguntade PG, Abiodun BJ, Ajayi AE, Van de Giesen N. Effetcs of charcoal production on soil physical properties in Ghana. Journal of Plant Nutrition and Soil Science. 2008;**171**:591-596

[6] Mukherjee A, Lal R. Biochar impacts on soil physical properties and greenhouse gas emissions. Agronomy. 2013;**3**:313-339

[7] Major J, Lehmann J, Rondon M. Biochar application to a topical oxisol modifies water relations. In: Poster at the 2006 World Soil Congress; Philadelphia, PA. 2006

[8] Chen HX, Du ZL, Guo W, Zhang QZ. Effects of biochar amendment on cropland soil bulk density, cation exchange capacity, and particulate organic matter content in the North China plain. Ying Yong Sheng Tai Xue Bao. 2011;**22**(11):2930-2934

[9] Kinney TJ, Masiello CA, Dugan B, Hockaday WC, Dean MR, Zygourakis K, et al. Hydrologic properties of biochars produced at different temperatures. Biomass and Bioenergy. 2012;**41**:34-43. DOI: 10.1016/j.biombioe.2012.01.033

[10] Jeffery S, Verheijen FGA, van der Velde M, Bastos AC. A quantitative review of the effects of biochar application to soils on crop productivity using meta-analysis. Agriculture, Ecosystems & Environment. 2011;**144**:175-187

[11] Mukherjee A, Lal R. The biochar dilemma. Soil Research. 2014;**52**:217-230

[12] Tryon EH. Effect of charcoal on certain physical, chemical, and biological properties of forest soils. Ecological Monographs. 1948;**18**:81-115. DOI: 10.2307/1948629

[13] Masiello C, Dugan B, Brewer C, Spokas K, Novak J, Liu Z. Biochar effects on soil hydrology. In: Biochar for Environmental Management Science, Technology and Implementation. 2nd ed. United Kingdom: Routledge; 2015

[14] Downie A, Crosky A, Munroe P. Physical properties of biochar. In: Lemman J, Joseph S, editors. Biochar for Environmental Management: Science and Technology. London, UK: Earthscan; 2009. pp. 13-32

[15] Lua AC, Yang T, Guo J. Effects of pyrolysis conditions on the properties of activated carbons prepared from pistachio-nut shells. Journal of Analytical and Applied Pyrolysis. 2004;**72**:279-287

[16] Boateng AA. Characterization and thermal conversion of charcoal derived from fluidized-bed fast pyrolysis oil

production of switch grass. Industrial Engineering and Chemical Research. 2007;**46**:8857-8862

[17] Byrne CE, Nagle DC. Carbonized wood monoliths – Characterization. Carbon. 1997;**35**:267-273

[18] Brown RA, Kercher AK, Nguyen TH, Nagle DC, Ball WP. Production and characterization of synthetic wood chars for use as surrogates for natural sorbents. Organic Geochemistry. 2006;**37**:321-333

[19] Emmerich FG, Sousa JC, Torriani IL, Luengo CA. Applications of a granular model and percolation theory to the electrical resistivity of heat treated endocarp of babassu nut. Carbon. 1987;**25**:417-424

[20] Liu Z, Dugan B, Masiello CA, Gonnermann HM. Biochar particle size, shape, and porosity act together to influence soil water properties. PLoS One. 2017;**12**(6):e0179079. DOI: 10.1371/journal.pone.0179079

[21] Zimmerman AR. Abiotic and microbial oxidation of laboratory-produced black carbon (biochar). Environmental Science & Technology. 2010;**44**:1295-1301

[22] Hammes K, Torn MS, Lapenas AG, Schmidt MWI. Centennial black carbon turnover observed in a Russian steppe soil. Biogeosciences. 2008;**5**:1339-1350

[23] Dexter AR. Soil physical quality. Part I. Theory, effects of soil texture, density, and organic matter, and effects on root growth. Geoderma. 2004;**120**:201-214

[24] Dexter AR, Czyz EA. Soil physical quality and the effects of management practices. In: Wilson MJ, Maliszewska-Kordybach B, editors. Soil Quality, Sustainable Agriculture and Environmental Security in Central and Eastern Europe. NATO Science Series 2, Environmental Security. Vol. 69. Dordrecht: Kluwer Academic Publishers; 2000. pp. 153-165

[25] Troeh FR, Thompson LM. Soils and Soil Fertility. Iowa, US: Blackwell Publishing; 2005

[26] Chan KY, Van Zwieten L, Meszaros I, Downie A, Joseph S. Agronomic values of green-waste biochar as a soil amendment. Australian Journal of Soil Research. 2007;**45**:629-634

[27] Laird DA, Fleming P, Davis DD, Horton R, Wang BQ, Karlen DL. Impact of biochar amendments on the quality of a typical midwestern agricultural soil. Geoderma. 2010;**158**:443-449

[28] Githinji L. Effect of biochar application rate on soil physical and hydraulic properties of a sandy loam. Archives of Agronomy and Soil Science. 2014;**60**:457-470

[29] Jones BEH, Haynes RJ, Phillips IR. Effect of amendment of bauxite processing sand with organic materials on its chemical, physical and microbial properties. Journal of Environmental Management. 2010;**91**:2281-2288

[30] Scott HL, Ponsonby D, Atkinson CJ. Biochar: An improver of nutrient and soil water availability? What is the evidence? CAB Reviews. 2014;**9**(019):1-19

[31] Gaskin JW, Speir A, Morris LM, Ogden L, Harris K, Lee D, Das KC. Potential for pyrolysis char to affect soil moisture and nutrient status of loamy sand soil. Proceedings of the Georgia Water Resources Conference; University of Georgia, USA. 2007

[32] Omondi MO, Xia X, Nahayo A, Liu X, Korai PK, Pan G. Quantification of biochar effects on soil hydrological properties using meta-analysis of literature data. Geoderma. 2016;**274**:28-34

[33] Hansen V, Hauggaard-Nielsen H, Petersen CT, Mikkelsen TN, Müller-Stöver D. Effects of gasification biochar on plant-available water capacity and plant growth in two contrasting soil types. Soil and Tillage Research. 2016;**161**:1-9

[34] Brewer CE, Chuang VJ, Masiello CA, Gonnermann H, Gao X, Dugan B, et al. New approaches to measuring biochar density and porosity. Biomass and Bioenergy. 2014;**66**:176-185

[35] Waqas M, Li G, Khan S, Shamshad I, Reid BJ, Qamar Z, et al. Application of sewage sludge and sewage sludge biochar to reduce polycyclic aromatic hydrocarbons (PAH) and potentially toxic elements (PTE) accumulation in tomato. Environmental Science and Pollution Research. 2015;**22**(16):12114-12123

[36] Wang D, Zhang W, Hao X, Zhou D. Transport of biochar particles in saturated granular media: Effects of pyrolysis temperature and particle size. Environmental Science and Technology. 2013;**47**:821-828

[37] Wu L, Pan L, Mitchell J, Sanden B. Measuring saturated hydraulic conductivity using a generalized solution for single-ring infiltrometers. Soil Science Society of America Journal. 1999;**63**:788-792

[38] Deveraux RC, Sturrock CJ, Mooney SJ. The effects of biochar on soil physical properties and winter wheat growth. Earth and Environmental Science Transactions of the Royal Society of Edinburgh. 2012;**103**:13-18

[39] Barnes RT, Gallagher ME, Masiello CA, Liu Z, Dugan B. Biochar-induced changes in soil hydraulic conductivity and dissolved nutrient fluxes constrained by laboratory experiments. PLoS One. 2014;**9**:e108340. DOI: 10.1371/journal.pone.0108340

[40] Asai H, Benjamin SK, Haefele SM, Khamdok S, Koki H, Yoshiyuki K, et al. Biochar amendment techniques for upland rice production in Northern Laos: 1. Soil physical properties, leaf SPAD and grain yield. Field Crops Research. 2009;**111**:81-84

[41] Major J, Rondon M, Molina D, Riha SJ, Lehmann J. Nutrient leaching in a Colombian savanna oxisol amended with biochar. Journal of Environmental Qualilty. 2012;**41**:1076-1086

[42] George C, Wagner M, Kucke M, Rillig MC. Divergent consequences of hydrochar in the plant-soil system: Arbuscular mycorrhiza, nodulation, plant growth and soil aggregation effects. Applied Soil Ecology. 2012;**59**:68-72

[43] Busscher WJ, Novak JM, Ahmedna M. Physical effects of organic matter amendment of a southeastern us coastal loamy sand. Soil Science. 2011;**176**:661-667

[44] Busscher WJ, Novak JM, Evans DE, Watts DW, Niandou MAS, Ahmedna M. Influence of pecan biochar on physical properties of a Norfolk loamy sand. Soil Science. 2010;**175**:10-14

[45] Peng X, Ye LL, Wang CH, Zhou H, Sun B. Temperature- and duration-dependent rice straw-derived biochar: Characteristics and its effects on soil properties of an ultisol in southern China. Soil and Tillage Research. 2011;**112**:159-166

[46] Brodowski S, John B, Flessa H, Amelung W. Aggregate-occluded black carbon in soil. European Journal of Soil Science. 2006;**57**:539-546

[47] Mukherjee A. Physical and chemical properties of a range of laboratory-produced fresh and aged biochars [Ph.D. thesis]. Gainesville, FL, USA: University of Florida; 2011

Chapter 3

Increasing the Amount of Biomass in Field Crops for Carbon Sequestration and Plant Biomass Enhancement Using Biochar

Saowanee Wijitkosum and Thavivongse Sriburi

Abstract

The agricultural sector, especially in developing countries, is vulnerable to the effects of climate change partially caused by greenhouse gas (GHG) emissions from agricultural areas. Field crops are capable of bio-sequestration in its aboveground and belowground biomass. Incorporating biochar as a soil amendment increases its potential to become an important bio-sequestration which makes the agricultural sector a key contributor to climate change mitigation. This chapter discussed and presented data obtained from research on biochar using to increase plant biomass for carbon sequestration purposes. The biochar was produced from cassava stems by pyrolysis using a patented retort that was especially designed for agriculturalists to produce a low-cost biochar for their own use. The ability to increase biomass of field crops for carbon sequestration is crucial towards reducing the GHG emissions. This research also shed light on an innovative agricultural method, in comparison to traditional farming, that leads to sustainable agriculture in the long run. The biochar research is also a way to transfer research knowledge from laboratory to practical use.

Keywords: biochar, carbon sequestration, carbon storage, biomass, agriculture

1. Introduction

The agricultural sector contributes to climate change problems through greenhouse gas (GHG) emission from various agricultural activities. However, the agricultural sector is also a carbon sink, both in terms of its potential to store carbon in various forms and its cultivated area, where agricultural areas are scattered all over the globe. Thus, agricultural areas could potentially be utilized as effective carbon sequestration areas. Moreover, the Food and Agriculture Organization (FAO) of the United Nations (UN) has also suggested the use of agricultural areas for carbon sequestration to reduce GHG emissions [1, 2].

According to the UN Framework Convention on Climate Change (UNFCCC), the measurement of GHG emission reduction and the measurement of carbon capture and storage in agricultural sectors should not have any effect on food production and farmers. The framework has been specially emphasized in agricultural and developing countries, where most of the population are farmers and are from a low socioeconomic background. Therefore, GHG reduction can be performed in the form of a carbon sink in agricultural areas, where the carbon that is sequestered by biomass during photosynthesis or bio-sequestration [2, 3] can reduce the amount of GHG emission throughout the plant's life time [4–7]. Bio-sequestration appears to be a suitable and viable means of mitigation for long-term climate objectives. Many research reports have suggested that plants are capable of bio-sequestration in the form of accumulated biomass in their stems and in the soil [1, 6, 8]. The notion of carbon sequestration in biomass as a means to climate change mitigation is based upon the aim of storing carbon in different types of forest areas [9–13]. Although carbon sequestration in plant biomass in agriculture is an effective tool for climate change mitigation, carbon sequestration in agricultural sectors has not yet been intensively evaluated in agricultural countries. The level of carbon sequestration in the aboveground and belowground biomass of plants depends on the plant's biomass and thus varies with the plant species/cultivar, age, and quantity of the plants [14, 15]. Some or many field crop areas are suitable for carbon sequestration without negative impacts on farmers and food production.

Biochar is a highly stable substance that is high in fixed carbon. Incorporating it into agricultural soils has the potential to become an important means for GHG reduction. Biochar contributes to GHG reduction by retaining the carbon within the soils and within the plants or bio-sequestration [16–20]. Moreover, biochar has been widely used as a soil amendment to improve crop yields, in terms of the quantity and quality [21–24]. It also improves the physical, chemical, and biological characteristics of the soil [23, 25–28]. Therefore, using biochar as a soil amendment can help reduce requirements for agrochemical fertilizers, which is one of the causes of GHG emissions. It fits within the framework from the UNFCCC and Kyoto Protocol report [29, 30].

In this context, this chapter discussed and presented data obtained from research on biochar using to increase plant biomass for carbon sequestration purposes. The biochar was produced from feedstocks by pyrolysis using a patented retort that was especially designed for agriculturalists to produce a low-cost biochar for their own use. The biochar research is also a way to transfer research knowledge from laboratory to practical use.

2. Biochar, carbon sequestration, and plant biomass relationships

The indirect storage of carbon is the natural CO_2 storage system from the growth of plants, which is an inexpensive method and can be implemented anywhere in the world. Most of the time, it is implemented in forested areas; however, according to a number of research studies, agricultural areas as well as forested areas are considered a promising place to store carbon [2, 4–7, 23]. It could reduce greenhouse gases as well as perform as a sink of agricultural CO_2. Undoubtedly, the method is given considerable attention, especially by the Food and Agriculture Organization (FAO) who gives very much importance on measures to reduce greenhouse gases [31]. The movement of carbon and the variation scale of CO_2 from air to soil increase carbon in soil. Subsequently, there is a decreasing amount of CO_2 released from soil to air. Therefore, carbon storage is an influential mechanism that tremendously affects the reduction of greenhouse gases, which has approximately 89% of technical efficiency, whereas there was a 9% and 2% reduction of methane gas and nitrous oxide released from soil, respectively. Moreover, the movement of carbon from carbon emissions to carbon absorptions would efficiently reduce the variation of the atmosphere [32].

IPCC [1] characterized carbon storage in forested areas into five places including biomass above ground, underground biomass, dead trees, and organic carbon in the soil, all of which consist of storage in trees, and most of it is stored underground. Each type of trees possesses different carbon storage efficiency and accumulated carbon according to the wood and types of wood changing according to the present related conditions [33–35], such as the age of the forest, the type of the forest, and the tree sizes [36], the forest density [37], the forest structure [38], and more. Nevertheless, plants except big trees can be adopted in storing carbon with more studies concerning the amount of carbon absorption or the amount of carbon storage in the life cycle of each plant. Carbon would be captured since the initial growth of plants until their full maturity. After plants are fully grown, the captured carbon would remain stable. Carbon indirect storage adopts photosynthesis of the plants, which depends on CO_2 to propel the chemical reaction to water turning into glucose and oxygen, as in Eq. (1).

$$6CO_2 + 6H_2O \xrightarrow[\text{Sunlight energy}]{} C_6H_{12}O_6 + 6O_2 \tag{1}$$

Carbon storage in the soil of agricultural and forested areas is an approach several countries have adopted to reduce GHG emissions. The implementation could be immediate and inexpensive, relying on the photosynthesis of plants that store carbon in the plant tissues (cores, leaves, fruits, and roots). After the death of these parts, these organic parts decompose, while it is also hard for some parts to decompose such as humus, which remain in the soil as organic matters. The number of the fallen plant components varies according to habitats of living organisms. The factors that affect the fallen plants include plant types, environment, the care of the plants, and duration. By and large, products obtained from the plants are more than fallen plants, possibly attributable to the plant age compared to the plant density [14]. According to that, biochar is adopted in the carbon storage in the soil in order to cut the cycle of being released to the atmosphere. Furthermore, methane and nitrous oxide emissions could be cut down from agricultural areas; hence, this process is effective in greenhouse gas reduction.

Biochar can improve the degraded soil, which has been proved by research to effectively enhance agricultural products, increasing the biomass of plants [23, 39–41], which is an indirect way to reduce greenhouse gases (Carbon Negative Technology) [17, 18, 42]. What is more, biochar has a high volume of fixed carbon. After the process of pyrolysis, there would be only 50% of carbon left in biochar [18, 44, 45]. Carbon in biochar is steady and hard to decompose by microorganisms in the soil, making biochar remain underground for a long time. Thus, this could be considered a way of carbon storage in the soil [20, 46], different from other organic matters such as plants, green manure, compost manure, and manure. These could decompose quickly, especially in tropical areas, giving rise to a high volume of CO_2 emissions in a rapid manner [47]. For this reason, agricultural areas with the integration of biochar can store carbon more effectively than those with the integration of biomass with the same amount of carbon [48]. According to the research study by Maraseni [49], once there is a change in the agricultural areas from enlargement by deforestation and slash and burn systems to deforestation and slash and char systems, there is 12% reduction of losing carbon. Biochar made of grass could reduce 3 tons of CO_2 emissions per 1 ton of biochar [50].

3. Pilot project for biochar application for sustainable agriculture in Thailand

3.1 Study area

The study of increasing biomass in feeding maize (*Zea mays* L.) was performed on experimental plots in Pa Deng-Biochar Research Center (Pd-BRC), Pa Deng sub-district, Kaeng Krachan district, Petchaburi province, Thailand. This is part of the Huay Sai Royal Development Study Centre. The topology is undulating and rolling. The soil is sandy loam with a medium to high soil permeability, a medium to very low organic matter (OM = 0.04–1.16), and a pH that ranges from slightly alkaline to extremely acidic. The land has very low soil fertility and experiences soil erosion and water scarcity [51]. The majority of the area in Pa Deng is a slope complex with a gradient of more than 35%. Therefore, the Pa Deng area is enclosed by hills that limit the land utilization to only 12% of the total area [52]. The low soil fertility and limited area available for agriculture lead to the heavy use of agrochemicals among farmers to improve the quality and yield of their agricultural products. This creates long-term negative effects on the soil and environment.

3.2 Research design and experimental plots

A completely randomized design was used for this study. There were 7 treatments each with 4 replications giving a total of 16 experimental plots. Each experimental plot was 3 × 5 m in size. The maize was planted in two crop cycles. After harvesting the first cycle, the treatments were left in their original condition with no further addition of biochar or organic fertilizer. The maize was planted in May and was harvested in August. Pa Deng has been suffering from droughts for a long period of time. The crops were planted during the absence of rain period and in the strong sunlight. The crops were watered from water sprinklers.

There are seven treatments in total. Four treatments consisted of soil plus 5.6 ton/ha of organic fertilizer with different amounts of added biochar at 0 (TBC0), 5 (TMBC0.5), 25 (TMBC2.5), and 30 (TMBC3.0) ton/ha, respectively. The other three treatments consisted of soil plus added biochar at 0, 5 (TBC0.5), 25 (TBC2.5), and 30 (TBC3.0) ton/ha, respectively. TBC0 was the controlled treatment.

The organic fertilizer used in this study was produced from the composting of soybean stems, and its characteristics were as follows: pH 8.3, electrical conductivity (EC) of 3.50 dS/m, 40.30 wt.% OM, 23.43 wt.% total organic carbon (TOC), 1.70 wt.% total nitrogen (total N), 0.87 wt.% total phosphorus (total P_2O_5), 3.54 wt.% total potassium (total K_2O), and a 13.75 C/N ratio. In general, all the properties of fertilizer were shown in **Table 1**. The organic fertilizer used in this study was in accordance with all the parameters of the Organic Fertilizer Standard of the Thai Department of Agriculture in 2005 [53].

The maize used in this study was a single-cross hybrid CP 888 variety (flint corn) with strong stems. This maize can be waited for a long harvest. The maize is drought tolerant and can grow well in upland areas with medium precipitation making it suitable in the Pa Deng area. It is also popular among farmers. Biochemical pesticides and herbicides were used to prevent pests and weeds, especially during the period of 13–25 days after seeding emergence. This is the most critical period to prevent flora and pests from severely affecting the crops [53, 54].

3.3 Biochar production and its characteristic

Biochar was produced from cassava stems (cassava crop waste) by pyrolysis process using the Controlled Temperature Biochar Retort for Slow Pyrolysis Process (patented) that the research team invented to suit local uses. The biochar process is simple and low-cost [20, 23]. The retort was a controlled temperature biochar retort for slow pyrolysis which was complied with the standard set by FAO [56], with a controlled temperature between 450 and 600°C. After the process was finished, the biochar was ground and sieved to less than 3 mm diameter. This particle size was selected since it improves soil aeration and other processes in the soil [55, 57].

Parameters	Units	Soil	Fertilizer	Cassava Biochar
pH	-	6.95 ± 0.19	8.30	9.60
OM	%	1.32 ± 0.18	40.30	25.89
OC	%	-	23.43	-
EC	dS/m	0.08 ± 0.01	3.50	1.35
CEC	$cmol_c/kg$	7.12 ± 0.43	-	11.00
Total N	%	0.09 ± 0.01	1.70	0.98
Avail. P	mg/kg	21.80 ± 5.20	-	-
Total P_2O_5	%	-	0.87	0.82
Exch. K	mg/kg	215.75 ± 16.76	-	-
Total K_2O	%	-	3.54	1.68
Physical properties				
- Surface area	m^2/g	-	-	200.46
- Total pore volume	cm^3/g	-	-	0.12
- Average pore diameter	Å	-	-	24.4
Composition of biochar				
- TC	%	-	-	58.46
- TOC	%	-	-	58.46
- H	%	-	-	2.24
- O	%	-	-	33.44
- H/C_{org} Ratio	molar	-	-	0.43

Table 1.
The properties of pre-experimental soil, fertilizer, and cassava biochar.

The biochar sampling method was adapted from the Standardized Product Definition and Product Testing Guidelines for Biochar that is used in soil [58] by collecting samples from every pyrolysis process. The samples were randomly selected from the ground biochar and analyzed for their specific surface area, total pore volume, average pore diameter, pH, EC, cation exchange capacity (CEC), OM, total carbon (C), total organic carbon (TOC), %hydrogen (H), %Oxygen (O), and the molar hydrogen to total organic carbon ratio (H/C_{org} Ratio).

The cassava biochar composites were comprised of 58.46 wt.% total C and 58.46 wt.% TOC. The biochar from the cassava stems had a specific surface area of 200.46 m^2/g, total pore volume of 0.12 cm^3/g and average pore diameter of 24.4 Å, with an alkaline pH of 9.6, EC of 1.35 dS/m, and CEC of 11.00 cmol/kg. The cassava biochar had a very high OM content of 25.89%, total N of 0.98%, total P_2O_5 of 0.82%, and total K_2O of 1.68% (**Table 1**).

The cassava stem biochar was high in carbon, mostly in the form of amorphous carbon in which the carbon atoms were attached in aromatic rings [18, 21, 22, 42, 44]. This chemical property makes the carbon in cassava stem biochar very stable [59–61] and creates a highly porous carbon structure in the biochar [60, 62]. The pyrolysis biochar at 450–600°C also contributed to the high stability of carbon [60, 63, 64]. The high porosity of biochar allows biochar to absorb and retain water and nutrients within the soil [23, 42, 55, 61, 65]. This helps with aeration and reduces soil density [18, 60, 66–68]. Moreover, the appropriate temperature during the pyrolysis process of the cassava stems also increased porosity on the biochar's surface which led to increased ions on the its surface [17, 18, 62, 69, 89]. This resulted in a high ion exchange capacity and high CEC [26, 42, 60, 69, 70]. As a result, the cassava stem biochar had a high capacity to retain and adsorb organic carbon and non-organic matters within the soil. Moreover, it also increased activities in the soil and ion exchange between nutrients in the form of soil solution.

Cassava biochar has high alkalinity (pH 9.6). Alkalinity affects the type of biomass made into biochar [25, 71, 72]. Moreover, biochar from cassava stems also had a high OM (25.9 wt.%), which would contribute to an increased OM level in the soil and improve the soil fertility. These physical and chemical characteristics and chemical formations in biochar made it suitable as a soil amendment to increase plant growth [23, 25, 43, 44, 55, 60, 74, 75] and soil amelioration in acidic soils.

3.4 Soil properties and soil character analysis

The soil in the experimental plots was analyzed before planting the crops. Soil was selected at random from areas scattered throughout each plot and taken from 0 to 30 cm depth. The samples were considered as composite samples in the soil analysis. Physical and chemical characteristics of the soil samplings were analyzed using the methods developed by the Soil Survey Staff [76], including the pH, OM (Walkley and Black method), soil texture (hydrometer method), CEC (leaching method), EC, total N (Kjeldahl method), available phosphorus (avail. P) (Bray II determine by spectrophotometer), and exchangeable potassium (exch. K) (ammonium acetate extraction determine by atomic absorption spectrophotometer).

The pre-experimental soil analysis results (**Table 1**) revealed that the soil in the experimental plots was a slightly alkaline sandy clay loam (%Sand = 57.0, %Silt = 22.5, %Clay = 20.5) with a pH of 6.95 and EC of 0.08 dS/m. It is suitable for growing flint corn for feeding animals [53]. The soil had a high level of primary macronutrients except total N (total N = 0.09%, avail. P = 21.80 mg/kg, and exch. K = 215.75 mg/kg) (**Table 1**).

The soil in this region had a very low fertility with an OM of 1.32%. The OM in soils is decomposed by soil microbes, and it depended on the carbon distribution at different soil densities, which helped prevent the decomposition [77].

3.5 Evaluation of the maize biomass

During the harvesting period, the maize was uprooted from the soil and washed with water. The plants were then left to dry in the shade before being measured for their whole plant fresh (wet) weight (FW). The plants were then cut so as to separate the roots, upper roots (stems + leaves + staminate), pods, and seeds. The FW of each part of the plant was measured then cut into small pieces and put in an oven at 70°C for 48 h or until the weight was stable (dry weight: DW). Using the FW/DW ratio, the crop biomass was estimated. After that, the DW of the plants was used to derive the moisture content (wt.%), from which the biomass in different parts of the crop in each experimental plot was calculated, derived from Eqs. (2) and (3):

$$\text{Biomass} = 100\,[\text{DW (g)}]/(\text{moisture content} + 100) \tag{2}$$

$$\text{Moisture content} = 100\,[\text{FW (g)} - \text{DW (g)}]/\text{FW (g)} \tag{3}$$

3.6 Analysis of carbon sequestration from maize grown in the different biochar-supplemented soils

The amount of carbon sequestered in each part of maize in the different experimental treatment plots consisted of the carbon concentration of the plant biomass, as shown in Eq. (4). The plant carbon stock was estimated by multiplying the total plant biomass with the carbon concentration (%). This study applied the FAO method [78] for carbon stock in biomass, derived from Eqs. (4) and (5):

$$\text{Biomass C} = [\text{Carbon concentration (\%) x biomass}]/100 \tag{4}$$

$$\text{Biomass } C_{\text{stock total}} = \text{Biomass } C_{ag} + \text{Biomass } C_{bg} \tag{5}$$

Biomass $C_{stock\ total}$ is the total stock of C in the biomass from every part of maize. The constituents of the biomass carbon stock aboveground were the carbon content in the upper roots, corn cobs, and seeds, while belowground they were the carbon content in the roots.

All the data collected from the different experiments and field samples during the study were compiled and processed for statistical analysis by analysis of variances (ANOVA). Comparisons between means were tested for significance with Tukey's multiple comparison test using the Statistical Package of the Social Science (SPSS) software. Significance was accepted at the $p < 0.05$ level.

3.7 Biomass of maize grown in the different biochar-supplemented soils

Biomass assessment during the first crop cycle (CC1) (**Figure 1**) indicated that the total biomass in the maize grown in TMBC3.0 was the highest (17.63 ton/ha), while the biomass was lowest (14.71 ton/ha) in the soil added fertilizer (TBC0). However, these numerical differences in the total biomass were not significant among all seven soil types. Comparing the results between biochar-incorporated treatments, it was apparent that the amount of biomass increased in relation to the amount of added biochar (highest in TBC3.0 and lowest in TBC0.5) and increased further if the fertilizer was also added. However, soil incorporated with fertilizer and the least amount of biochar (TMBC0.5) yielded less biomass than soil incorporated with solely biochar at the highest amount (TBC3.0), but again these differences were not statistically significant (**Figure 1**).

Maize biomass in the second crop (CC2) yielded (**Figure 1**) similar results to those of CC1, where numerically the highest total biomass was found in TMBC3.0, both in the whole plant (17.31 ton/ha) and in each part of the maize. Compared to the control, the total biomass and biomass of roots in TMC3.0 treatment showed significant results whereas the other ones did not. Even though there was no significant difference in biomass (total and each plant part) among soil types, which may reflect the low sample size relative to the level of intra-sample variation,

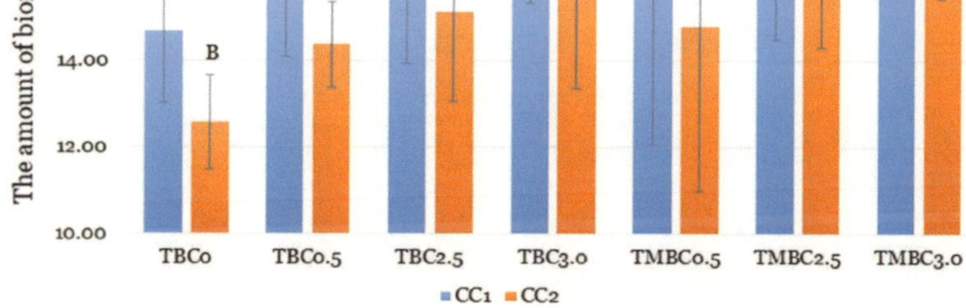

Figure 1.
*Total biomass in the maize grown in soil supplemented with different biochar levels for two successive crop cycles. CC1 and CC2 are the first and second crop cycles, respectively. Data are shown as the mean ± 1SD, derived from ** independent samples. Means within a row (small letter), or within a column (capital letter) between CC1 and CC2 of a given maize part, with a different letter are significantly different (p < 0.05).*

numerically it was apparent that incorporating the appropriate amount of biochar within the soil could increase the amount of biomass in every part of the maize.

Comparing between the two successive crop cycles (**Figure 2**), the amount of biomass found in each treatment in CC2 was less than in CC1, except for the roots in TBC2.5, TBC3.0, TMBC2.5, and TMBC3.0 that had a slightly higher biomass (0.061, 0.049, 0.120, and 0.125 ton/ha, respectively) in CC2 than in CC1. However, TMBC3.0, which received the highest amount of biochar plus fertilizer, had the least difference between the two crop cycles (−0.317 ton/ha) that the total biomass in the maize grown in TMBC3.0 was the highest in both crop cycles, while TBC0 (control) had the highest difference between the two crop cycles (−2.13 ton/ha). Thus, increasing the level of biochar in the soil (within this range of 5 to 3 ton/ha) numerically decreased the loss of biomass yield between the first and second successive cultivation. However, none of these numerical differences were statistically significant.

From the results, considering only the maize seed biomass that can be sold for animal feed, adding the fertilizer with highest amount of biochar into the soil gave the highest (yield) weight of maize seeds in both the first and second maize plantations, and adding only biochar into the soil gave a higher maize seed biomass in both crop cycles than that obtained when only adding fertilizer to the soil. The weight of maize seed biomass from TMB3.0 was the highest (6.280 ton/ha in CC1 and 6.149 ton/ha in CC2), while the results reported by Wijitkosum [55] revealed that TMB2.5 (13 cobs) had the highest average number of cobs per plant from 8 sample plants per treatment followed by TMB3.0 (12 cobs). In the second crop, the soil amendment with biochar and fertilizer still gave a high yield of maize seeds with only a small decrease in the biomass compared to that in the first crop cycle.

The increase of maize biomass obtained from the soil with added biochar reflects the high porosity, surface area, and ion exchange capacity of biochar [20, 21, 23, 44, 61, 62]. In addition, the highly aromatic chemical structure of

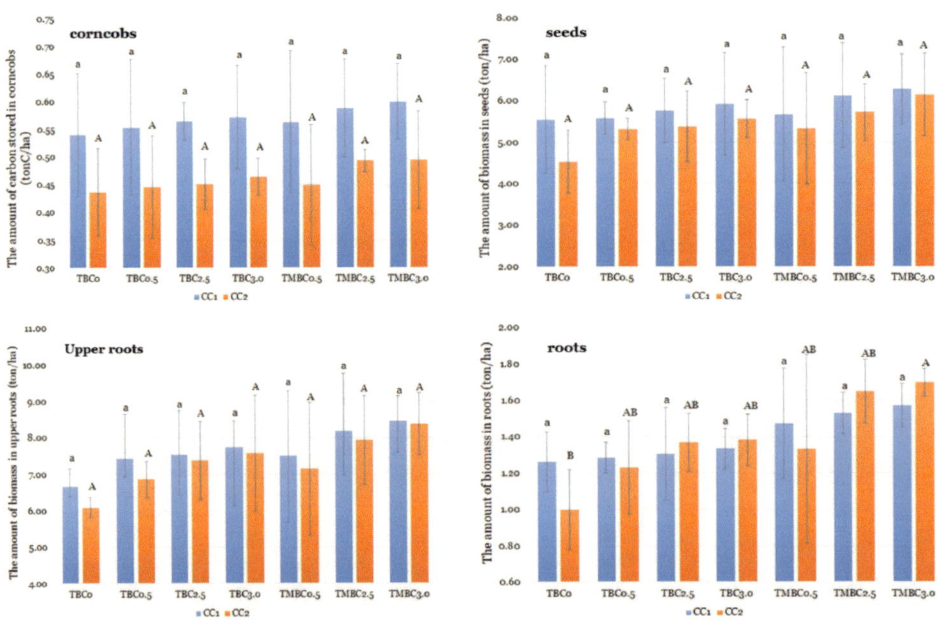

Figure 2.
*Biomass in each part of the maize grown in soil supplemented with different biochar levels for two successive crop cycles. CC1 and CC2 are the first and second crop cycles, respectively. Data are shown as the mean ± 1SD, derived from ** independent samples. Means within a row (small letter), or within a column (capital letter) between CC1 and CC2 of a given maize part, with a different letter are significantly different ($p < 0.05$).*

biochar leads to a high chance of oxidation reactions to form functional groups, and so biochar has many anions on its surface and hence a high ion exchange capacity [20, 42, 44, 65, 72, 73]. Moreover, biochar has many micropores that can absorb nutrients and anions from the soil solution [46, 59–62, 65, 79, 80] and to reduce nutrient leaching and provide a sustainable release to the plants.

The organic matter, important as a source of nutrients for maize growth, mostly came from the added fertilizer and some from the biochar and soil. Together, they support the growth of the roots and aid in absorbing more nutrients and transfer to the stem. The root biomass was increased in every soil amendment with biochar alone or with biochar and fertilizer, at all levels of biochar, and was higher than that obtained in the soil with only fertilizer added. This result gave the consistent with many studies (e.g. [20, 60, 72, 81, 82]) indicating that biochar could also contribute to the suitable environment for the growth of plant root. In the second maize plantation, the root biomass was significantly higher in all the biochar treatments, and especially for the addition of fertilizer with the highest level of biochar, than that obtained from the soil with only fertilizer added.

When the plant's roots grow well, they can absorb nutrients and water to build up the biomass in other parts of the plant. For example, potassium affects the growth, photosynthesis, carbohydrate synthesis, and leaf and seed formation [83–86]. Calcium affects the strength of the maize plant and activates development of the roots and leaves, as well as controlling the soil's pH [20, 87]. Biochar produced from cassava has a high nutrient content, reflected in the observation that maize grows well with a higher biomass when grown in soil with added fertilizer and biochar or added biochar compared to that in soil with only added fertilizer.

3.8 The amount of carbon sequestered from growing maize

The carbon stock in biomass in CC1 showed that the highest amount of carbon stored in biomass in TMBC3.0 at 7.22 ton/ha, while the lowest in TBC0 at 5.83 ton/ha (**Figure 3**). The study showed that the carbon storage in maize biomass was increased depending on the amount biochar added into the soil, especially when the biochar was added with the fertilizer. However, the carbon storage obtained with the

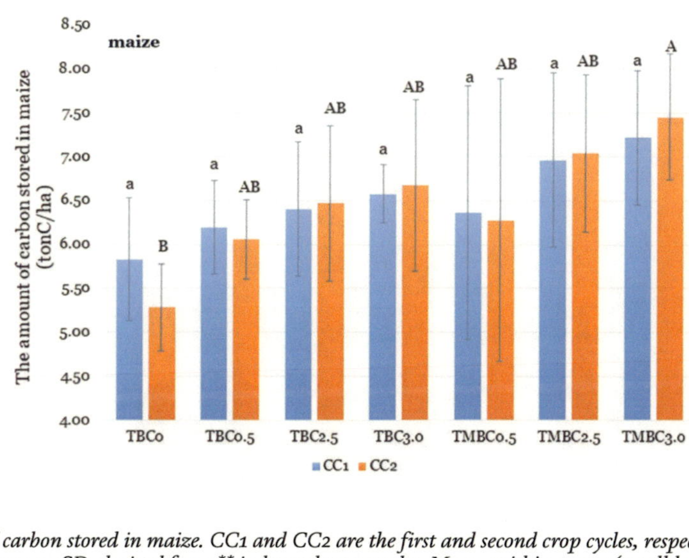

Figure 3.
*The amount of carbon stored in maize. CC1 and CC2 are the first and second crop cycles, respectively. Data are shown as the mean ± 1SD, derived from ** independent samples. Means within a row (small letter), or within a column (capital letter) between CC1 and CC2 of a given maize part, with a different letter are significantly different (p < 0.05).*

lowest ratio of biochar with fertilizer (TMB0.5) was lower than that in the biochar only treatment when sufficient biochar was added (TBC2.5 and TBC3.0). Carbon storage in each part of the maize and the total amount of carbon storage were not significantly different among the seven treatments. The highest percentage of carbon storage in the maize biomass was found in the upper roots (46.72–49.21%), followed by that in the seeds (33.71–35.69%), corncobs (8.32–9.27%), and roots (8.04–9.10%) (**Figures 4** and **5**).

With respect to the results from the CC2 (**Figure 3**), TMBC3.0 still gave the highest carbon storage (7.46 ton/ha), followed by TMBC2.5, TBC3.0, TBC2.5, TMBC0.5, TBC0.5, and TBC0. The amount of carbon storage was clearly different among the soil treatments, especially with the addition of fertilizer plus a high level of biochar which resulted in a significantly higher amount of carbon storage than the addition of fertilizer alone, which is the standard agricultural soil amendment used by farmers. Soil amendment with fertilizer and a sufficient amount of biochar (TMBC2.5 and TMBC3.0) resulted in significantly higher root carbon storage than the addition of only fertilizer to the soil. Similarly, the ratio of carbon storage in the other parts of the maize plants was in the same pattern as that seen in the first crop (**Figures 4** and **5**), being highest in the upper roots (46.50–48.21%), then the seeds (35.39–37.49%), corncobs (6.64–8.27%), and roots (7.57–9.55%).

With respect to the amount of carbon storage between the first and second maize plantings, the total carbon storage on maize was increased only in the soil treatments with sufficient biochar addition alone or with the fertilizer adding sufficient biochar. Treatment TMB3.0 gave the highest amount of carbon storage in maize (+0.235 ton/ha), followed by TBC3.0 (+0.094 ton/ha), TBC2.5 (+0.083 ton/ha), and TMBC2.5 (+0.076 ton/ha. In contrast, soil amendment without any biochar, but with the fertilizer only (TBC0), resulted in the highest level of decreased carbon storage (−0.551 ton/ha) between the two maize planting cycles.

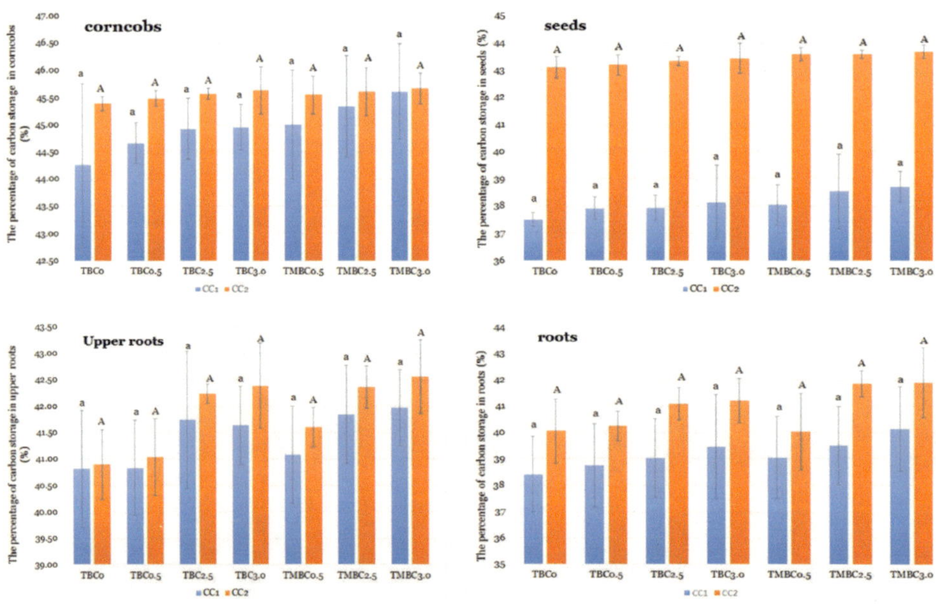

Figure 4.
*The percentage of carbon storage in different parts of maize. CC1 and CC2 are the first and second crop cycles, respectively. Data are shown as the mean ± 1SD, derived from ** independent samples. Means within a row (small letter), or within a column (capital letter) between CC1 and CC2 of a given maize part, with a different letter are significantly different ($p < 0.05$).*

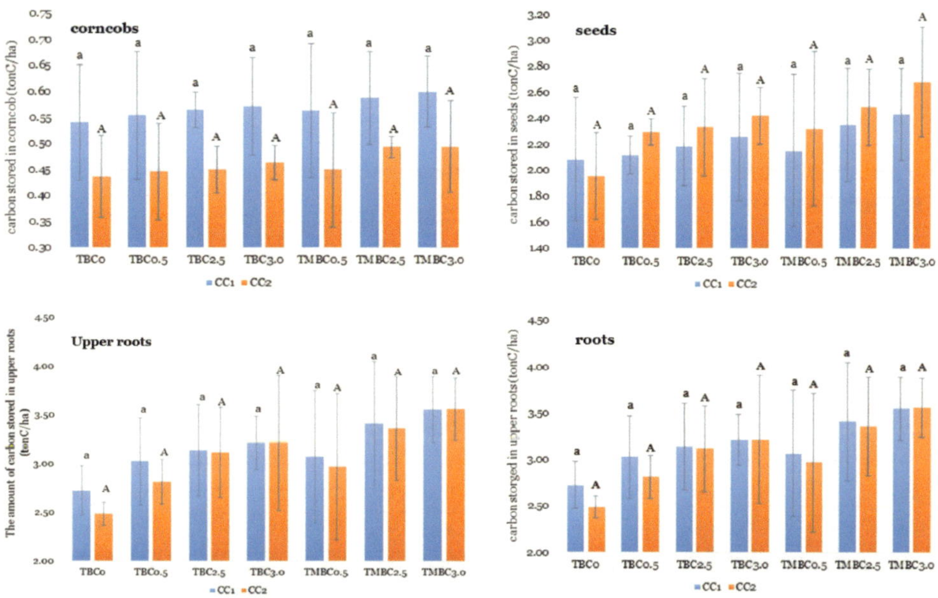

Figure 5.
*The amount of carbon stored in different parts of maize. CC1 and CC2 are the first and second crop cycles, respectively. Data are shown as the mean ± 1SD, derived from ** independent samples. Means within a row (small letter), or within a column (capital letter) between CC1 and CC2 of a given maize part, with a different letter are significantly different ($p < 0.05$).*

Considering the rate of total carbon change in maize biomass, the use of fertilizer (5.6 ton/ha) and biochar (30 ton/ha) (TMBC3.0) increased the amount of carbon storage in the maize biomass compared to that in the first crop cycle by 3.25%. The use of fertilizer alone (TBC0) or biochar alone showed a 9.45% or 2.28% decrease, respectively, in the total carbon storage in the second maize crop, whereas the soil amendment with fertilizer plus the lowest amount of biochar (TMBC0.5) gave only a 1.32% decrease in the total carbon storage in the maize biomass in the second crop.

Adding the appropriate amount of biochar into the soil promotes plant growth [23, 25, 55], especially the roots stems, leaves, stamen, and corn stalk, leading to an increased plant biomass. Moreover, the presence of biochar in the soil promotes the plant growth and productivity even without soil amendment with fertilizer because biochar is organic carbon that cannot be easily digested by soil microorganisms [17, 42, 59–61, 88]. Although the soil mixed with fertilizer initially provides sufficient nutrients for maize growth, this may be insufficient in the longer term for successive crops due to the rapid microbial degradation and leaching of the nutrients, leading to the requirement for continual reapplication of fertilizer every crop cycle. To help restore the soluble nutrients and reduce their leaching from soil, [21, 41, 45, 46, 89–91], especially in tropical regions where the soil has a low organic matter and high washout rate, the biochar with the fertilizer was applied. Under these conditions, adding organic matter alone to tropical soil is not stable in the long term because the soil has a low anion exchange capacity, and so much of soluble fertilizer is washed out before being absorbed by plant roots. Instead, the requirement to continuously add a high amount of organic matter to the soil increases the production cost and decreases the soil quality and environment in the long term [47, 57, 92, 94–95]. In contrast, when adding biochar with the fertilizer into the soil, the biochar helps improve both the physical and chemical properties of the soil allowing the plant's roots to absorb the nutrients over a longer time period [20, 42, 43, 60], and so the maize received enough nutrients continuously leading to higher productivities. Thus, the total biomass of the maize in second plantation in TMBC3.0 and TMBC2.5 had decreased by less than 10%.

4. Impact of biochar on biomass, bio-sequestration, and carbon sequestration

The massive and deep rooting systems in annual crops allow for direct movement of C into the soil and make it less available for removal by harvest [96]. Therefore, the results suggested that the incorporation of the appropriate amount of biochar into soil may help increase the amount of biomass in the maize. These results are in accordance with other biochar research, where the appropriate amount of biochar induced chemical reactions within the soil which enhanced the quantity and quality of the crops [23, 25, 28, 57, 98–100]. Incorporating biochar with the fertilizer could enhance and sustain the biomass gain from the fertilizer addition. Moreover, biochar remains in the soil for a long period of time with less leaching, and so it is not necessary to add more biochar every new crop cycle. The result from the main component (70–90% by weight) of biochar is amorphous carbon [23, 25, 43, 59] arranged in aromatic rings that are highly stable in the soil for long times [21, 22, 43, 59, 61]. Moreover, other important qualities of biochar are its high density of micropores, high surface area, and high ion exchange capacity. Therefore, biochar has good soil amendment qualities and can increase the agricultural productivity in terms of both the quality and quantity of crop obtained [10, 17, 20, 23, 25, 27, 28, 62, 91, 93, 97, 99].

The amount of biomass has a direct effect on the amount of carbon stored in the biomass. The quantity of biomass is an important source of replenishing organic carbon in the soil. The potential for soils to sequester C depends on the rate of biomass production relative to that exported, such as by microbial activity [96, 100]. The treatments that resulted in a high maize biomass also had a high amount of carbon in their biomass. Using biochar in agricultural areas had a positive impact on the maize and increased the amount of biomass stored in every part of the maize (roots, stems, leaves, tassels, seeds, and corncobs), as reported previously [23]. This is because the characteristics of biochar are beneficial for plants and its ability to be used for soil amelioration [70, 71, 101, 102].

The structure of biochar is amorphous, in the form of aromatic hydrocarbons bound with oxygenated functional groups, which influences its high stability characteristic [18–22, 42–44, 49, 70]. Moreover, its highly porous structure contains a large amount of micropores with a high surface area giving a high adsorption capacity for cations [65, 70, 72, 73, 75, 89–91, 99]. Therefore, incorporating biochar within the soil in agricultural areas benefits the soil ecosystem and the physical, biological, and chemical characteristics of the soil [17, 18, 22, 23, 25–28, 62, 73, 79, 80, 101, 102]. The soil becomes more fertile, which in turn leads to higher maize productivity. Maize grown in biochar-incorporated soils had a higher amount of carbon stored in every part of the plant.

5. Conclusion

A single application of biochar to the soil used for maize plantations significantly increased the carbon storage in the plants (biomass quantity and amount of carbon in the biomass) even in the second crop. The amount of carbon storage was further increased when the fertilizer was also added with the biochar to the soil.

The amount of plant biomass depends on the completion of plant growth, which is affected by the soil richness and nutrient availability. Adding organic material helps to improve the soil qualities and accelerate plant growth, but, especially in tropical soils, it can be washed out easily. The addition of biochar into the soil directly improves the physical and chemical properties of the soil, promotes microorganism activities and reduces nutrient leaching, and so leads to better plant growth and a higher biomass in the long term.

Carbon is stored in the soil directly by adding biochar, with its high stable carbon content, and will indirectly be the increased plant biomass. This is hence a method to reduce the carbon dioxide, a GHG emission, in agricultural areas and so help to mitigate climate change. This study revealed that adding a high amount of biochar together with fertilizer to agricultural soil only once is sufficient for at least two crops of maize and so would not only increase carbon storage in plants, but also the reduced fertilizer application will further reduce GHG release in agricultural areas and also reduce the production cost for farmers.

Acknowledgements

This research was supported by the "Minimizing GHG Emissions from Industrial and Agricultural Sectors to Reduce Adverse Impacts of Climate Change in Thailand, Sub Project: Reducing GHG Emission from Agricultural Sector by Using Biochar" funded by the 2014 In-depth Strategic Research Fund, Ratchadapisek Sompoch Endowment Fund, Chulalongkorn University. Furthermore, this chapter was also partially supported by "Building a Smart Community for Climate Change and Natural Disasters Adaptation. Sub-project: Using biochar in urban farming areas for food security and carbon sequestration on high-rise buildings" (CU59-002-IC), the 2016 Ratchadapisek Sompoch Endowment Fund for in-depth high potential research projects.

Author details

Saowanee Wijitkosum[1]* and Thavivongse Sriburi[2]

1 Environmental Research Institute, Chulalongkorn University, Bangkok, Thailand

2 Chula UNISEARCH, Chulalongkorn University, Bangkok, Thailand

*Address all correspondence to: w.m.saowanee@gmail.com

References

[1] Intergovernmental Panel on Climate Change. 2006 IPCC Guidelines for National Greenhouse Gas Inventories: Volume 4 Agriculture, Forestry and Other Land Use. IPCC [Internet]. 2006. Available from: http://www.ipcc-nggip.iges.or.jp/public/2006gl/vol4.html [Accessed: 2017-07-13]

[2] Lal R. Soil carbon sequestration impacts on global climate change and food security. Science. 2004;**304**:1623-1627. DOI: 10.1126/science.1097396

[3] Soil Science Society of America. Glossary of Soil Science Terms 2008. Wisconsin: American Society of Agronomy; 2008. 92 p

[4] Redondo-Brenes A, Montagnini F. Growth, productivity, aboveground biomass, and carbon sequestration of pure and mixed native tree plantations in The Caribbean Lowlands of Costa Rica. Forest Ecology and Management. 2006;**232**:168-178. DOI: 10.1016/j.foreco.2006.05.067

[5] Gorte R. Carbon Sequestration in Forests. CRS Report for Congress. Washington D.C.: Congressional Research Service; 2009. 22 p. DOI: https://fas.org/sgp/crs/misc/RL31432.pdf

[6] Intergovernmental Panel on Climate Change. Good Practice Guidance for Land Use, Land-Use Change and Forestry. Kanagawa: Institute for Global Environmental Strategies; 2003. 590 p

[7] Chavan BL, Rasal GB. Carbon sequestration potential of young Annona reticulate and Annona squamosa from University Campus of Aurangabad. International Journal of Physical and Social Sciences. 2012;**2**(3):193-198

[8] Food and Agricultural Organization. FAO NFMA – Support to Developing Countries on National Forest Monitoring and Assessment [Internet]. 2005. Available from: http://www.fao.org/forestry/19906-0fccd6930d9f093755955ace14354c928.pdf [Accessed 2017-07-10]

[9] Schimel DS, House JI, Hibbard KA, Bousquet P, Ciais P, Peylin P, et al. Recent patterns and mechanisms of carbon exchange by terrestrial ecosystems. Nature. 2001;**414**:169-172

[10] Peichl M, Arain MA. Above- and belowground ecosystem biomass and carbon pools in an age-sequence of temperate pine plantation forests. Agricultural and Forest Meteorology. 2006;**140**:51-63. DOI: 10.1016/j.agrformet.2006.08.004

[11] Lorenz K, Lal R. Biogeochemical C and N cycles in urban soils. Environment International. 2009;**35**:1-8. DOI: 10.1016/j.envint.2008.05.006

[12] Loaiza Usuga JC, Rodríguez Toro JA, Ramírez Alzate MV, de Jesús Lema Tapias A. Estimation of biomass and carbon stocks in plants, soil and forest floor in different tropical forests. Forest Ecology and Management. 2010;**260**:1906-1913

[13] Thomas SC, Martin AR. Carbon content of tree tissues: A synthesis. Forests. 2012;**3**:332-352. DOI: 10.3390/f3020332

[14] Lichaikul N. Change of soil carbon stock and sequestration after conversion of forest to reforestation and agricultural lands [Master's thesis]. Bangkok: King Mongkut's University of Technology Thonburi; 2004

[15] Jenkins J, Chojnacky D, Heath L, Birdsey R. National-scale biomass estimators for United States tree species. Forest Science. 2003;**49**(1):12-35

[16] Freibauer A, Rounsevell MDA, Smith P, Verhagen J. Carbon

sequestration in agricultural soils of Europe. Geoderma. 2004;**122**:1-23

[17] Lehmann J, Gaunt J, Rondon M. Bio-char sequestration in terrestrial ecosystems - A review. Mitigation and Adaptation Strategies for Global Change. 2006;**11**(2):403-427

[18] Novak JM, Busscher WJ, Laird DL, Ahmedna M, Watts DW, Niandou M. Impact of biochar amendment on fertility of a southeastern Coastal Plain soil. Soil Science. 2009;**174**(2):105-112

[19] Qian L, Chen L, Joseph S, Pan G, Li L, Zheng J, et al. Biochar compound fertilizer as an option to reach high productivity but low carbon intensity in rice agriculture of China. Carbon Management. 2014;**5**:145-154. DOI: 10.1080/17583004.2014.912866

[20] Yooyen J, Wijitkosum S, Sriburi T. Increasing yield of soybean by adding biochar. Journal of Environmental Research and Development. 2015;**9**(4):1066-1074

[21] Lehmann J, Rillig MC, Thies J, Masiello CA, Hockaday WC, Crowley D. Biochar effects on soil biota - A review. Soil Biology and Biochemistry. 2011;**43**(9):1812-1836

[22] Zhang HJ, Dong HZ, Li WJ, Zhang DM. Effects of soil salinity and plant density on yield and leaf senescence of field-grown cotton. Journal of Agronomy and Crop Science. 2012;**198**(1):27-37. DOI: 10.1111/j.1439-037X.2011.00481.x

[23] Wijitkosum S, Kallayasiri W. The use of biochar to increase productivity of indigenous upland rice (*Oryza sativa* L.) and improve soil properties. Research Journal of Pharmaceutical, Biological and Chemical Sciences. 2015;**6**:1326-1336

[24] Wijitkosum S, Mattayom B, Sriburi T. The effect of biochar on increasing the yield of seeda tomatoes. In: International Conference on Solid Wastes 2015: Knowledge Transfer for Sustainable Resource Management; 19-23 May 2015; Hong Kong. Hong Kong: ISWA; 2015

[25] Sriburi T, Wijitkosum S. Biochar amendment experiments in Thailand: Practical examples. In: Bruckman VJ, Varol EA, Uzun BB, Liu J, editors. Biochar: A Regional Supply Chain Approach in View of Climate Change Mitigation. Cambridge: Cambridge University Press; 2016. pp. 351-367

[26] Lehmann J, Gaunt J, Rondon M. Biochar sequestration in terrestrial ecosystems - a review. Mitigation and Adaptation Strategies for Global Change. 2006;**11**:403-427

[27] Yamato M, Okimori Y, Wibowo IF, Anshori S, Ogawa M. Effects of the application of charred bark of *Acacia mangium* on the yield of maize, cowpea and peanut, and soil chemical properties in South Sumatra, Indonesia. Journal of Soil Science and Plant Nutrition. 2006;**52**:489-495

[28] Masulili A, Utomo WH, Syechfani MS. Rice husk biochar for rice based cropping system in acid soil 1. The characteristics of rice husk biochar and its influence on the properties of acid sulfate soils and rice growth in west Kalimantan, Indonesia. The Journal of Agricultural Science. 2010;**2**(1):39-47

[29] Green R, Tobin B, O'Shea M. Above and below ground biomass measurements in an unthinned stand of Sitka spruce (*Picea sitchensis* (Bong) Carr.). European Journal of Forest Research. 2007;**126**:179-188

[30] Almgir M, Al-amin M. Regeneration status in a proposed biodiversity conservation area of Bangladesh. Proceedings of the Pakistan Academy of Sciences. 2007;**3**:165-172

[31] Food and Agricultural Organization. WRB Map of World Soil Resources, 2002 updates [Internet]. 2002. Available from: http://www.fao.org/ag/agl/agll/wrb/soilres.stm [Accessed: 2017-07-10]

[32] United State Geological Survey. Carbon Sequestration (workshop 1999) [Internet]. 2006. Available from: http://edcintl.cr.usgs.gov/carbonseq/workshop.html [Accessed: 2006-12-02]

[33] Ogawa H, Yoda K, Ogino K, Kira T. Comparative ecological studies on three main type of forest vegetation in Thailand II. Plant Biomass Nature and Life in Southeast Asia. 1965;**4**:49-80

[34] Tangtham N, Tantasirin C. An assessment of policies to reduce carbon emissions in the Thai forestry sector with emphasis on forest protection and reforestation for conservation. In: Proceedings of FORTROP'96 International Conference; 25-28 November 1996; Bangkok. pp. 100-121

[35] Negi JDS, Manhas RK, Chauhan PS. Carbon allocation in different components of some tree species of India: A new approach for Carbon estimation. Current Science. 2003;**85**(11):1528-1531

[36] Terakunpisut J, Gajaseni N, Ruankawe N. Carbon sequestration potential in aboveground biomass of Thong Pha Phum national forest, Thailand. Applied Ecology and Environmental Research. 2007;**2**:93-102

[37] Senpaseuth P, Navanugraha C, Pattanakiat S. The estimation of carbon storage in dry evergreen and dry dipterocarp forests in Sang Khom District, Nong Khai Province, Thailand. Environment and Natural Resources Journal. 2009;**7**(2):1-11

[38] Kaewkrom P, Kaewkla N, Thummikkapong S, Punsang S. Evaluation of carbon storage in soil and plant biomass of primary and secondary mixed deciduous forests in the lower northern part of Thailand. African Journal of Environmental Science and Technology. 2011;**5**(1):8-14

[39] Winsley P. Biochar and bionenergy production for climate change. New Zealand Science Review. 2007;**64**:5-10

[40] Ueno M, Kawamitsu Y, Komiya Y, Sun L. Carbonisation and gasification of bagasse for effective utilisation of sugarcane biomass. International Sugar Journal. 2007;**110**:22-26

[41] Bridgewater AV. Biomass for energy. Journal of the Science of Food and Agriculture. 2006;**86**:1755-1768

[42] Glaser B, Lehmann J, Zech W. Ameliorating physical and chemical properties of highly weathered soils in the tropics with charcoal: A review. Biology and Fertility of Soils. 2002;**35**:219-230

[43] Lehmann J, Joseph S. Biochar: Environmental Management. London: Earthscan; 2009. 448 p

[44] Lehmann J, Czimczik C, Laird D, Sohi S. Stability of biochar in soil. In: Lehmann J, Joseph S, editors. Biochar for Environmental Management: Science and Technology. London: Earthscan; 2009. pp. 169-182

[45] Liang B, Lehmann J, Sohi SP, Thies JE, O'Neill B, Trujillo L, et al. Black carbon affects the cycling of non-black carbon in soil. Organic Geochemistry. 2010;**41**:206-213

[46] Glaser B, Wiedner K, Seelig S, Schmidt HP, Gerber H. Biochar organic fertilizers from natural resources as substitute for mineral fertilizers. Agronomy for Sustainable Development. 2015;**35**(2):667-678

[47] Steiner C. Slash and char as alternative to slash and burn - Soil charcoal amendments maintain soil

fertility and establish a carbon sink [thesis]. Bayreuth: University of Bayreuth; 2007

[48] Kwapinski W, Byrne CMP, Kryachko E, Wolfram P, Adley C, Leahy JJ, et al. Biochar from biomass and waste. Waste and Biomass Valorization. 2010;**1**:177-189

[49] Maraseni TN. Biochar: Maximizing the benefits. International Journal of Environmental Studies. 2010;**67**(3):319-327

[50] Roberts KG, Gloy BA, Joseph S, Scott NR, Lehmann J. Life cycle assessment of biochar systems: Estimating the energetic, economic, and climate change potential. Environmental Science & Technology. 2010;**44**:827-833

[51] Wijitkosum S. Impacts of land use changes on soil erosion in Pa Deng sub-district, adjacent area of Kaeng Krachan National Park, Thailand. Soil and Water Research. 2012;**7**:10-17

[52] Wijitkosum S. Critical factors affecting the desertification in Pa Deng, adjoining area of Kaeng Krachan national Park, Thailand. Environmental Asia. 2014;7:87-98

[53] Department of Agriculture. DOA Notice on the Standard of Organic Fertilisers B.E. 2548 (in Thai.) [Internet]. 2005. Available from: http://www.ratchakitcha.soc.go.th/DATA/PDF/2548/00172707.PDF [Accessed: 2015-04-03]

[54] Department of Agricultural Extension. Handbook for Maize Production (in Thai). Bangkok: Department of Agricultural Extension; 2008

[55] Wijitkosum S. The use of biochar for food security in maize production. Unisearch Journal. 2017;**4**(3):9-14

[56] Sriburi T. Testing properties of biochar from wood residues before using as a soil amendment. In: The Proceedings of a Conference on Natural Resource Management and Quality of Life Improvement under the Royal Initiation, Huay Sai Royal Development Study Center, Petchaburi, Thailand; 26 August 2011; Petchaburi. pp. 31-49

[57] Bruun EW. Application of fast pyrolysis biochar to a loamy soil effects on carbon and nitrogen dynamics and potential for carbon sequestration [thesis]. Roskilde: National Laboratory for Sustainable Energy; 2011

[58] International Biochar Initiative. Standardized Product Definition and Product Testing Guidelines for Biochar That Is Used in Soil [Internet]. 2014. International Biochar Initiative. Available from: http://www.biochar-international.org/characterizationstandard [Accessed: 2017-08-01

[59] Kim KH, Kim J, Cho T, Choi JW. Influence of pyrolysis temperature on physicochemical properties of biochar obtained from the fast pyrolysis of pitch pine (*Pinus rigida*). Bioresource Technology. 2012;**118**:158-162

[60] Sohi S, Elisa LC, Evelyn K, Roland B. Biochar, Climate Change and Soil: A Review to Guide Future Research. CSIRO Land and Water Science Report 05/09. 2009

[61] Downie A, Crosky A, Munroe P. Physical properties of biochar. In: Lehmann J, Joseph S, editors. Biochar for Environmental Management: Science and Technology. London: Earthscan; 2009. pp. 13-32

[62] Chen YQ, Yang HP, Wang XH. Biomass-based pyrolytic polygeneration system on cotton stalk pyrolysis: Influence of temperature. Bioresource Technology. 2012;**107**:411-418

[63] Steinbeiss S, Gleixner G, Antonietti M. Effect of biochar amendment on soil carbon balance and soil microbial activity. Soil Biology and Biochemistry. 2009;**41**:130-1310

[64] Sun Y, Gao B, Yao Y, Fang J, Zhang M, Zhou Y, et al. Effects of feedstock type, production method, and pyrolysis temperature on biochar and hydrochar properties. Chemical Engineering. 2014;**240**:574-578

[65] Atkinson C, Fitzgerald J, Hipps N. Potential mechanisms for achieving agricultural benefits from biochar application to temperate soils: A review. Plant and Soil. 2010;**337**:1-18

[66] Jones BEH, Haynes RJ, Phillips IR. Effect of amendment of bauxite processing sand with organic materials on its chemical, physical and microbial properties. Environmental Management. 2010;**91**:2281-2288

[67] Bhogal A, Nicholson FA, Chambers BJ. Organic carbon additions: Effects on soil bio-physical and physico-chemical properties. European Journal of Soil Science. 2009;**60**:276-286

[68] Hati KM, Swarup A, Dwivedi AK, Misra AK, Bandyopadhyay KK. Changes in soil physical properties and organic carbon status at the topsoil horizon of a vertisol of central India after 28 years of continuous cropping, fertilization and manuring. Agriculture, Ecosystems and Environment. 2007;**119**:127-134

[69] Schmidt MW, Noack AG. Black carbon in soils and sediments: Analysis, distribution, implications and current challenges. Global Biogeochemical Cycles. 2000;**14**:777-793

[70] Amonette JE, Joseph S. Characteristics of biochar: Microchemical properties. In: Lehmann J, Joseph S, editors. Biochar for Environmental Management: Science and Technology. London: Earthscan; 2009. pp. 33-52

[71] Liu L, Shen G, Sun M, Cao X, Shang G, Chen P. Effect of biochar on nitrous oxide emission and its potential mechanisms. Journal of the Air and Waste Management Association. 2014;**64**:894-902

[72] Zhang J, Liu J, Liu R. Effects of pyrolysis temperature and heating time on biochar obtained from the pyrolysis of straw and lignosulfonate. Bioresource Technology. 2015;**176**:288-291

[73] Steiner C. Slash and char as alternative to slash and burn - Soil charcoal amendments maintain soil fertility and establish a carbon sink [thesis]. Bayreuth: University of Bayreuth; 2007

[74] Basso B, Fiorentino C, Cammarano D, Cafiero G, Dardanelli J. Analysis of rainfall distribution on spatial and temporal patterns of wheat yield in Mediterranean environment. European Journal of Agronomy. 2012;**41**:52-65

[75] Laird DA, Fleming PD, Karlen DL, Wang B, Horton R. Biochar impact on nutrient leaching from a Midwestern agricultural soil. Geoderma. 2010;**158**:436-442

[76] Soil Survey Staff. Soil survey field and laboratory methods manual. Soil survey investigations report No. 51, Version 2.0. In: Burt, R, Soil Survey Staff, editors. Washington, DC: US Government Printing Office; 2014

[77] Bouajila A, Gallali T. Soil organic carbon fractions and aggregate stability in carbonated and no carbonated soils in Tunisia. Journal of Agronomy. 2008;**7**:127-137

[78] Ponce-Hernandez R, Koohafkan P, Antoine J. Assessing Carbon Stocks and Modelling Win-Win Scenarios of Carbon Sequestration through Land-Use Changes. Rome: Food and Agricultural Organization of the United Nations; 2004

[79] Schulz, H., Dunst, G. and Glaser, B. Positive effects of composted biochar on plant growth and soil fertility. Agronomy for Sustainable Development. 2013;**33**(4):817-827

[80] Kloss S, Zehetner F, Dellantonio A, Hamid R, Ottner F, Liedtke V, et al. Characterization of slow pyrolysis biochars: Effects of feedstocks and pyrolysis temperature on biochar properties. Journal of Environmental Quality. 2012;**41**:990-1000

[81] Khodake SP, Borale R, Petare RK. Ichthyofaunal diversity in Jamkhedi reservoir in Dhule district of Maharashtra, India. Journal of Environmental Research and Development. 2014;**9**(1):177-183

[82] Manoj B. Bio-processing of Indian coals by micro-organisms: An investigation. Journal of Environmental Research and Development. 2014;**9**(1):209-215

[83] Armstrong DL. Potassium for agriculture. Better Crops with Plant Food. 1998;**82**:4-5

[84] De Datta SK. Mineral and fertilizer management of rice. In: De Datta SK, Principles and Practices of Rice Production. New York: John Wiley; 1981. p. 348-419

[85] Hartt CE. Effects of potassium deficiency upon translocation of ^{14}C in attached blades and entire plants of sugarcane. Plant Physiology. 1969;**44**:1461-1469

[86] Johnston A, Steen I. Understanding Potassium and Its Use in Agriculture. Brussels: European Fertilizer Manufacturers Association; 2003. 40 p

[87] Korb N, Jones C, Jacobsen J. Potassium Cycling, Testing, and Fertilizer Recommendations. Nutrient Management Module Number 5. Montana: Montana state University Extension Service; 2005. 12 p

[88] Singh B, Singh BP, Cowie AL. Characterisation and evaluation of biochars for their application as a soil amendment. Australian Journal of Soil Research. 2010;**48**(7):516-525

[89] Chan KY, Van Zwieten L, Meszaros I, Downie A, Joseph S. Agronomic values of greenwaste biochar as a soil amendment. Australian Journal of Soil Research. 2007;**45**(8):629

[90] Lehmann J. Biological carbon sequestration must and can be a win-win approach. Climatic Change. 2009;**97**(3):459-463

[91] Liang B, Lehmann J, Solomon D, Kinyangi J, Grossman J, O'Neill B, et al. Black carbon increases cation exchange capacity in soils. Soil Science Society of America Journal. 2006;**70**:1719-1730

[92] Brady NC, Weil RR. The Nature and Properties of Soils. 13th ed. New Jersey: Pearson; 2002. 960 p

[93] Schulz, H. and Glaser, B. Effects of biochar compared to organic and inorganic fertilizers on soil quality and plant growth in a greenhouse experiment. Journal of Plant Nutrition and Soil Science. 2012;**175**(3):410-422

[94] Sparkes J, Stoutjesdijk P. Biochar: Implications for Agricultural Productivity. ABARES Technical Report 11.6, Canberra, Australia. Australian Bureau of Agricultural and Resource Economics and Sciences; 2011

[95] Tiessen H, Cuevas E, Chacon P. The role of soil organic matter in sustaining soil fertility. Nature. 1994;**371**:783-785

[96] Lemus R, Lal R. Bioenergy crops and carbon sequestration. Critical Reviews in Plant Sciences. 2005;**24**: 1-21

[97] Major J, Rondon M, Molina D, Riha SJ, Lehmann J. Maize yield and nutrition during 4 years after biochar application

to a Colombian savanna oxisol. Plant and Soil. 2010;**333**:117-128

[98] Butnan S, Deenik JL, Toomsan B, Antal MJ, Vityakon P. Biochar properties influencing greenhouse gas emissions in tropical soils differing in texture and mineralogy. Journal of Environmental Quality. 2016;**45**:1509-1519

[99] Asai H, Samson BK, Stephan HM, Songyikhangsuthor K, Homma K, Kiyono Y, et al. Biochar amendment techniques for upland rice production in Northern Laos 1. Soil physical properties, leaf SPAD and grain yield. Field Crops Research. 2009;**111**(1):81-84

[100] Williams RJ, Hutley LB, Cook GD, Russell-Smith J, Edwards A, Chen X. Assessing the carbon sequestration potential of mesic savannas in the Northern Territory, Australia: Approaches, uncertainties and potential impacts of fire. Functional Plant Biology. 2004;**31**:415-422

[101] Ibrahim H, Hatira A, Gallali T. Relationship between nitrogen and soil properties: Using multiple linear regressions and structural equation modelling. International Journal of Research and Reviews. 2013;**2**:1-7

[102] Mašek O, Brownsort PA. Research on Production of Bespoke Biochar. Poster Presented at the 2nd UK Biochar Research Centre Conference; 28-29 October 2010; United Kingdom. UKBRC: Rothamsted; 2010

Chapter 4

Influence of Sewage Sludge Biochar on the Microbial Environment, Chinese Cabbage Growth, and Heavy Metals Availability of Soil

Guangwei Yu, Shengyu Xie, Jianli Ma, Xiaofu Shang, Yin Wang, Cheng Yu, Futian You, Xiaoda Tang, Héctor U. Levatti, Lanjia Pan, Jie Li and Chunxing Li

Abstract

The effects of sewage sludge biochar (SSB) on the microbial environment, Chinese cabbage yield, and heavy metals (HMs) availability of soil were comprehensively investigated in this study. Results showed that the concentrations of the dehydrogenase (DHA) and urease in the soil added with 10% SSB were 3.60 and 1.67 times as high as that of the control soil, respectively, after planting; the concentrations of the bacteria, fungi, ammonia-oxidizing archaea (AOA), and ammonia-oxidizing bacteria (AOB) in the soil added with 10% SSB after planting reached 2.84, 2.62, 1.76, and 2.23 times, respectively, compared with those of the control group; the weights of the aboveground and underground parts of Chinese cabbage were 5.82 and 8.67 times as high as those of the control group, respectively. Moreover, the addition of SSB enhanced the immobilization of Cr, Ni, and Cd. All in all, SSB can improve the microbial environment of soil and inhibit the availability of HMs, which is very important for their utilization in barren soil.

Keywords: sewage sludge biochar, soil, Chinese cabbage, microbial environment, heavy metals

1. Introduction

Because of rapid economic development, more than 30 million tons of wet sewage sludge (SS) are produced in China every year [1]. SS contains lots of organic pollutants, microorganisms, eggs of parasitic organisms, and heavy metals (HMs), which makes it an obvious threat to ecological environment [2]. Conventional disposal technologies such as landfill, incineration, and agricultural application encounter many environmental problems; so, they cannot be widely used [3]. Especially, the direct application of SS in agricultural production is strictly banned due to the problem of pathogens and contaminants [4].

The pyrolysis of SS is a technology in which SS is heated under zero or low-oxygen condition to produce sewage sludge biochar (SSB) and pyrolysis oil and gas. After conversion into SSB, all the pathogens and organic pollutants in SS are eliminated and the volume of SS is significantly reduced [5]. Also, the oil and gas produced by pyrolysis can save the input of external energy as supplemental fuel [6]. Apart from the applications mentioned above, SSB has numerous special advantages in improving soil quality and crop growth. First of all, biochar possesses a porous structure that can influence the soil's structure, porosity, particle size distribution, and density, which contributes to increasing the soil water-holding capacity and microbial activity [7]. Furthermore, biochar is alkaline and can improve the pH of soil [8]. Finally, biochar is rich in plenty of nutrients such as nitrogen, phosphorus, potassium, etc., exhibiting a positive effect on plant growth [9]. Song et al. [10] studied the influence of pyrolysis temperature and proportion of SSB on garlic yield and HMs accumulation and found that the SSB produced at 450°C and its addition at 25% could improve the yield of garlic well and inhibit HMs accumulation in garlic. Khan et al. [4] investigated the effects of SSB on rice yield, HMs bioaccumulation, and greenhouse gas emission and found that SSB amendments increased the pH, total nitrogen, organic carbon, and available nutrients of soil and crop yield, and decreased HMs bioavailability and N_2O emission. In addition, there are a large amount of studies on the influence of SSB on plant growth and HMs migration that have proved the positive effects of biochar addition [11–13].

Based on the pilot-scale plant on pyrolysis of SS with capacity 30 t/d in Xiamen, and our previous studies, it was found that the HMs in SS were converted into a more stable state after hydrothermal pretreatment combined with pyrolysis and the obtained SSB could be used to prepare ceramsite [14–16]. However, the study of the influence of SSB from the pyrolysis of hydrothermally treated SS on the microbial environment of soil during planting is still indispensable. On the one hand, the soil microorganisms are involved in many biochemical processes, including the degradation and conversion of organic matter, the mineralization and immobilization of nutrients, and the formation and stabilization of soil aggregates [17]. On the other hand, the soil microorganisms are also a repository of soil nutrients and an important nutrition source for plant growth [18]. In this study, we chose the common and easy-to-grow Chinese cabbage as the planting crop to investigate the influence of SSB from the pyrolysis of hydrothermally treated SS on the physical and chemical properties and microbial environment of soil before and after planting. Furthermore, the growth status of Chinese cabbage and HMs availability were also studied.

2. Materials and methods

2.1 Materials

The used soil was collected from a farmland near an abandoned mine in Longyan, Fujian Province, China. The soil was sieved and homogenized after collection. SS was obtained from a wastewater treatment plant in Xiamen, China. Then, the SS was disposed via hydrothermal treatment at 160°C for 1 hour, and followed by filtration and pyrolysis by a rotary furnace at 500°C for 3 hours to obtain SSB in the pilot-scale plant in Xiamen, Fujian Province [19]. The high-quality and early raping NO.5 seed of Chinese cabbage was chosen as the testing plant.

2.2 Chinese cabbage pot experiment

The Chinese cabbage pot experiment was carried out in a greenhouse located in Xiamen, Fujian province, China (24.36 N–118.3 E) and the height and diameter of the polyethylene pot were 15 and 20 cm, respectively. To investigate the influence of SSB on the properties of soil, Chinese cabbage growth, and HMs availability, SSB was added with an SSB-to-soil mass ratio of 1:9 (10% SSB) in pot and the pure soil served as a control group. The total weight of soil or treated soil in each pot was 5.0 kg. Every pot experiment was assessed by four replicates. After seeding, each pot was treated with watering regularly and thinned out to ensure that only one Chinese cabbage grows. When the pot experiment finished (about 55 days), the soil and Chinese cabbage were collected to conduct relative tests, respectively.

2.3 Analysis methods

The pH was measured according to the agricultural trade standard of China (NY/T 1377-2007) and the solution was analyzed with a UB-7 pH meter (Ultra Basic, US). Electrical conductivity (EC) was measured according to the national environmental protection standard of China (HJ 802-2016) and the solution was analyzed with a Cond 3110 conductometer (Teltracon 325, Germany). Surface area was calculated by the Brunauer-Emmett-Teller (BET) method after testing using nitrogen adsorption/desorption isotherms with an apparatus (TriStar II 3020 V1.01, USA). Elemental analysis was conducted by an elemental analyzer (Vario MAX, Germany). The concentrations of nutrient elements were analyzed by digestion in an acid mixture [15] and the solution was determined by ICP-OES (Optima 7000DV, USA). The concentrations of available HMs in the sample were measured by the DTPA extraction method [20] and the solution was determined by ICP-MS (Agilent 7500cx, USA). The surface functional group of SSB was analyzed by FTIR spectrometry (iS10, Thermo, USA) and the morphology of SSB was analyzed by scanning electron microscopy (SEM, S-4800, Hitachi, Japan).

The dehydrogenase (DHA) activity in soil was measured by the triphenyltetrazolium chloride (TTC) spectrophotometric method [21]. The urease activity was measured by Nesslerization [22]. The molecular target genes of bacteria, fungi, ammonia-oxidizing archaea (AOA), and ammonia-oxidizing bacteria (AOB) were measured by quantitative real-time polymerase chain reaction (RT-PCR) analysis [23] and the information of primers is shown in **Table 1**. A standard curve was obtained by tenfold dilution of recombinant plasmid acquired in each molecular target gene of the above microorganisms and each sample was repeated three times. The SYBR® Premix Ex Taq™ kit from Bao Biological Engineering (Dalian, China) Co. Ltd. was used for analysis at Roche Lightcycler® 480 PCR. The quantitative PCR reaction system was 20 μL, including 1 μL of tenfold diluted DNA template, 10 μL of SYBR® Premix Ex Taq™, 0.2 μL (20 μM) of forward and reverse primers

Target gene	Primer name	Primer sequence (5′–3′)
Bacteria 16S rRNA	58F	CCTACGGGAGGCAGCAG
	517R	ATTCCGCGGCTGCTGGCA
Fungi 18S iRNA	ITS3	GCATCGATGAAGAACGCAGC
	ITS4	TCCTCCGCTTATTGATATGC
AOA amoA	Arch-amoAF	STAATGGTCTGGCTTAGACG
	Arch-amoAR	GCGGCCATCCATCTGTATGT
AOB amoA	amoA-1F	GGGGTTTCTACTGGTGGT
	amoA-2R	CCCCTCKGSAAAGCCTTCTTC

Table 1.
RT-PCR amplification primers.

respectively, and 8.6 μL of sterilized distilled water. The procedure of PCR consisted of denaturation at 95°C for 5 min, denaturation at 94°C for 30 s, annealing at 55°C for 45 min, and extension at 72°C for 1 min, followed by 40 cycles of denaturation, annealing, and extension at 72°C for 10 min.

3. Results and discussions

3.1 Basic properties of the original soil and SSB

The physical and chemical properties of the original soil and SSB are listed in **Table 2**. SSB has higher pH, EC, and BET surface area compared with the soil, which shows that the addition of SSB can improve the physicochemical properties of soil, such as pH, salinity content, water retention, the adsorption of nutrient, and microbial population [24]. In particular, the change of pH in soil indicates the occurrence of some chemical and biological reactions. The contents of C, H, N, and S in biochar depend on the feedstock and pyrolysis condition. The H/C and C/N ratios represent the aromaticity of biochar and the capacity for organics to release inorganic N [10, 25]. In this study, the H/C ratio of SSB is lower (<0.1) than that of the soil, which suggests that SSB has higher aromaticity and can exist in the soil for many years [25]. However, the higher C/N ratio of SSB inhibits the release of inorganic N compared with the original soil. In addition, SSB contains higher concentrations of K, Na, P, and Ca compared with the soil, which indicates that the addition of SSB can increase the fertility of soil.

The FTIR spectra of SSB is shown in **Figure 1a**. The identified bands are assigned to the stretching vibrations of hydroxyl functionalities (3446 cm^{-1}), amide bond stretching (1637 cm^{-1}), bending vibration of methyl group (1385 cm^{-1}), carbon-oxygen single bond in phenol (1186 cm^{-1}), and carbon-oxygen double bond (1050 cm^{-1}) [10, 25, 26]. In addition, the stretching vibrations between 600 and 800 cm^{-1} can be related to the aromatic and heteroaromatic compounds, and the bands below 600 cm^{-1} can be attributed to the organic and inorganic halogen compounds [25]. The SEM micrograph of SSB is shown in **Figure 1b**. There are lots of lumps and holes

Parameters	Soil	SSB
pH	5.32 ± 0.03	10.00 ± 0.04
EC (μS/cm)	203.67 ± 2.22	871.33 ± 3.78
Moisture (%)	0.26 ± 0.00	ND[a]
BET surface area (m^2/g)	0.51	13.05
Carbon (%)	3.08 ± 0.02	7.84 ± 0.02
Hydrogen (%)	1.04 ± 0.03	0.63 ± 0.03
Nitrogen (%)	0.26 ± 0.00	0.34 ± 0.00
Sulfur (%)	3.96 ± 0.04	3.82 ± 0.08
K (mg/g)	8.37 ± 0.05	20.33 ± 0.06
Na (mg/g)	0.86 ± 0.01	10.57 ± 0.05
P (mg/g)	1.51 ± 0.02	7.28 ± 0.05
Ca (mg/g)	0.03 ± 0.00	39.66 ± 0.11

[a]*ND, not detected.*

Table 2.
Physical and chemical properties of soil and SSB.

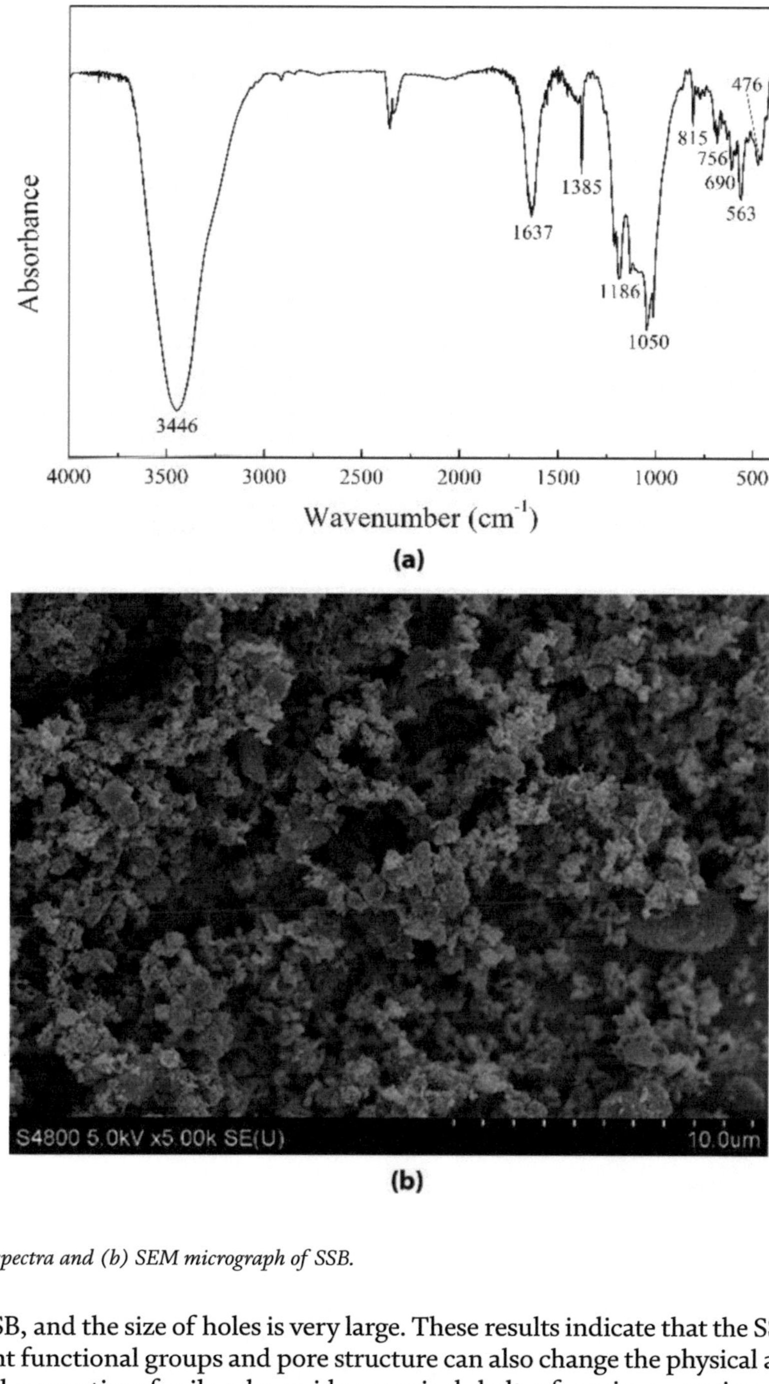

Figure 1.
(a) FTIR spectra and (b) SEM micrograph of SSB.

in the SSB, and the size of holes is very large. These results indicate that the SSB with abundant functional groups and pore structure can also change the physical and chemical properties of soil and provide a survival shelter for microorganism [27].

3.2 Effects of SSB addition on the physicochemical property of soil

The effects of SSB addition on the pH and EC of soil are shown in **Figure 2**. The pH of the control soil increased remarkably after planting, which indicated that the acid organic matter in soil was decomposed during Chinese cabbage planting [28]. Also, the addition of SSB adjusted the pH of soil from acidic to neutral and the pH increased from 7.12 to 7.49 after planting. **Figure 2b** shows that the EC of the control soil increased slightly after cabbage planting, but it is just 382 μS/cm and close to the

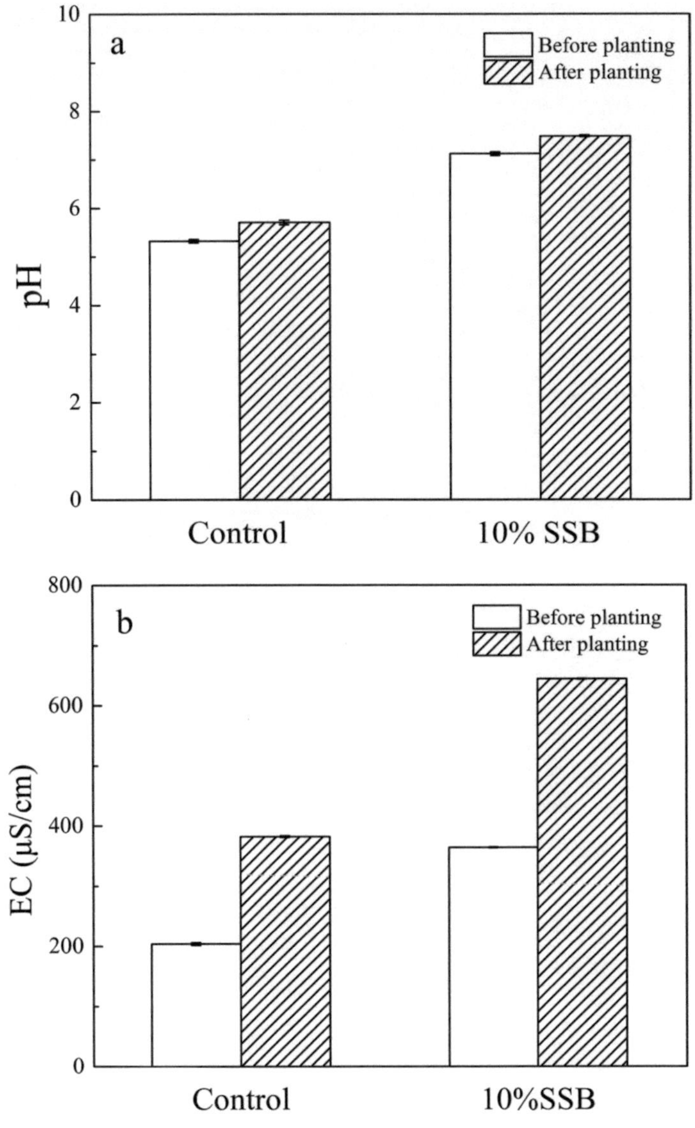

Figure 2.
Effects of SSB addition on the pH (a) and EC (b) of soil.

EC of the soil added with 10% SSB before planting. The EC of the soil with 10% SSB addition increased from 364 to 644 μS/cm after planting and the increase rate was 76.92%. When EC is lower than 500 μS/cm or higher than 2000 μS/cm, the phenomenon of lacking nutrient or seedling burning will occur during planting [29]. Therefore, adding SSB in soil could adjust the EC to a suitable range (500–2000 μS/cm) for plant growth. The above results are because a number of alkaline ions such as hydrocarbon anion, bicarbonate, carbonate, and phosphate in SSB were released during planting and increased the pH and EC of soil effectively [30, 31].

3.3 Effects of adding SSB on the microbiological property of soil

3.3.1 Effects of adding SSB on the DHA and urease in soil

DHA plays a key role in the decomposition process of organic matter and can be used as an indicator for the evaluation of total cell oxidation activity [32]. Therefore,

DHA activity is used to characterize the intensity of microbial activity. Urease can convert urea into ammonia and carbon dioxide or ammonium carbonate, and it reflects the intensity of nitrogen relevant reactions in the soil system [33]. The effects of SSB addition on the concentrations of DHA and urease in soil are shown in **Figure 3**. The addition of SSB increased the concentrations of DHA and urease in soil before planting, which rose from 3.83 μg IPTF/(g h) and 16.53 μg NH_3-N/(g h) to 14.33 μg IPTF/(g h) and 32.00 μg NH_3-N/(g h), respectively. Whether SSB is added or not, the concentrations of DHA and urease in soil increased after planting, and the concentrations of the DHA and urease in the soil added with 10% SSB reached 3.60 and 1.67 times as high as those of the control soil. These results implied that adding SSB could improve the activities of DHA and urease in soil, promote anaerobic microbial growth and synthesis of enzymes, and enhance microbial activity. This is because SSB influenced enzyme activity with the changes of physiochemical properties

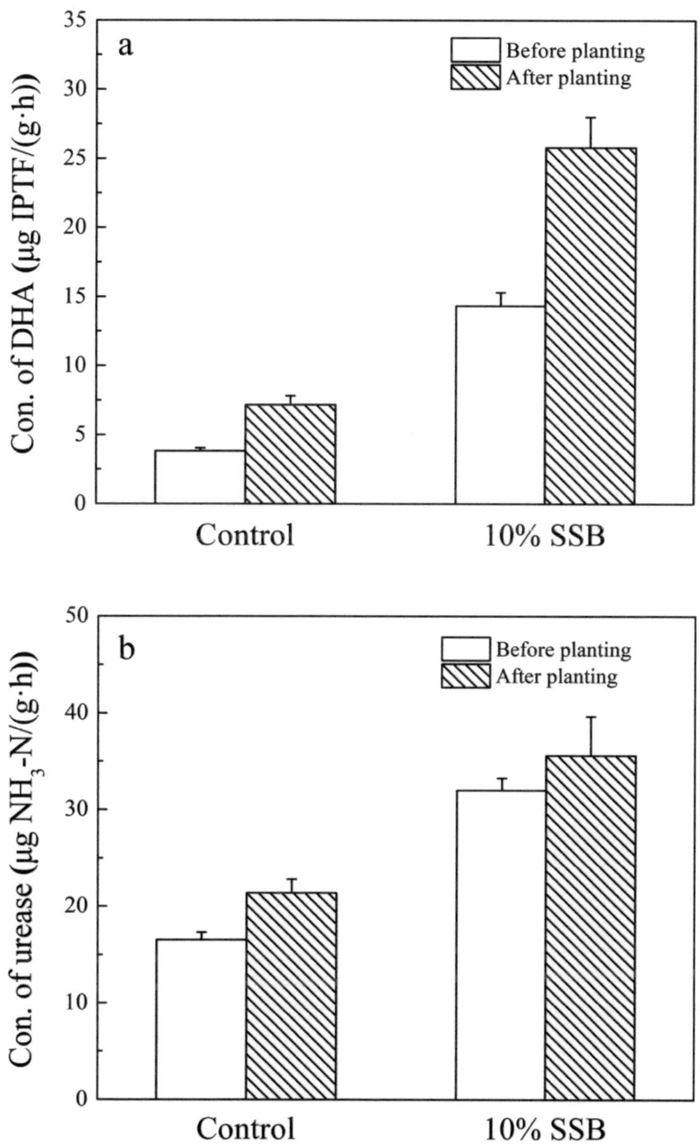

Figure 3.
Effects of SSB addition on the concentrations of DHA (a) and urease (b) in soil.

(especially pH) in soil, and the adsorption of enzymes and soil organic matter on SSB also changed the kinetic properties of enzyme activity [17].

3.3.2 Effects of SSB addition on bacteria and fungi in soil

In the planting process, bacteria play an important role in the transformation of organic and inorganic matter in soil, while fungi have significant effects on the carbon and energy cycle in soil [18]. The bacteria and fungi counts are important indicators of microbial activity intensity, and effectively reflect whether the environment of soil is suitable for crop growth or not. The effects of SSB addition on the concentrations of bacteria and fungi in soil are shown in **Figure 4**. The addition of SSB increased the concentrations of bacteria and fungi in soil before planting, which rose from 2.43×10^6 and 0.77×10^6 CFU/g to 20.60×10^6 and 3.67×10^6 CFU/g, respectively. Whether SSB is added or not, the concentrations of both bacteria and fungi in soil increased after planting, and the bacteria and fungi

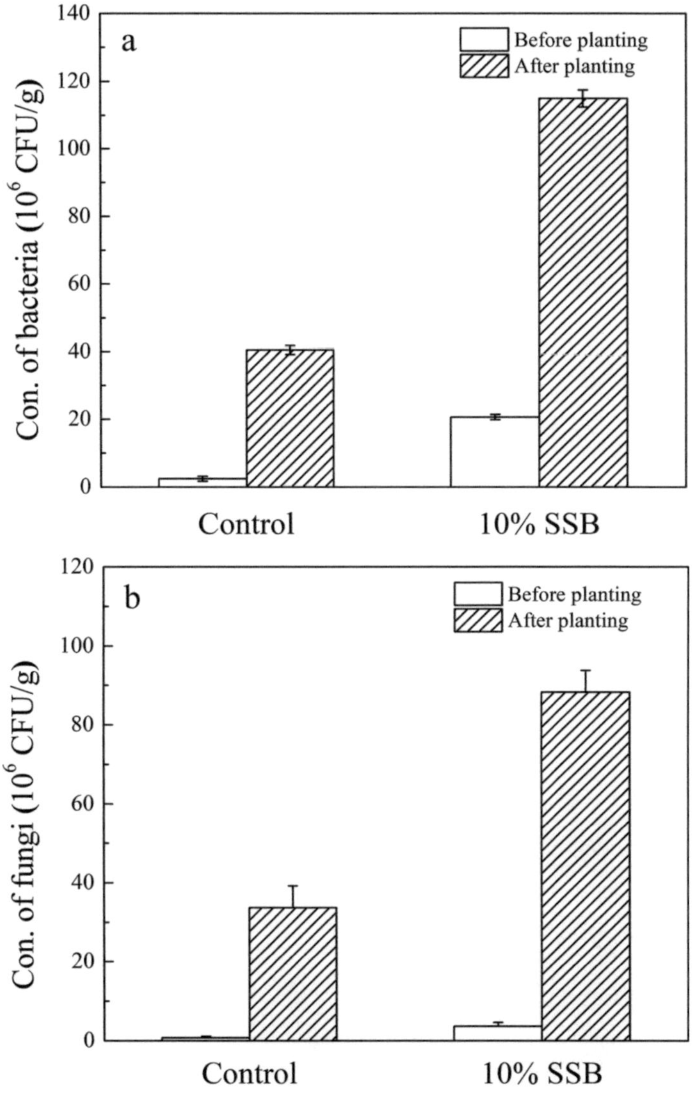

Figure 4.
Effects of SSB addition on the concentrations of bacteria (a) and fungi (b) in soil.

concentrations in soil added with 10% SSB reached 2.84 and 2.62 times as high as those of the control soil, respectively. These results showed that the addition of SSB had beneficial modulation effects on the concentrations of bacteria and fungi during planting, and it could effectively enhance the microbial property of soil.

3.3.3 Effects of adding SSB on the AOA and AOB in soil

AOA and AOB associated with the nitrification of soil are called the nitrifying bacteria. The higher concentrations of AOA and AOB can improve the conversion of other forms of nitrogen into available nitrogen fertilizer so as to enhance the fertility of soil and promote plant growth [34]. The effects of SSB addition on the concentrations of the AOA and AOB in soil are displayed in **Figure 5**. The addition of SSB increased the concentrations of AOA and AOB in soil before planting, which rose from 4.83×10^6 and 2.47×10^6 amoA copies/g to 8.63×10^6 and 6.07×10^6 amoA copies/g, respectively. Whether SSB is added or not, the concentrations of both AOA and AOB in soil increased after planting, and the AOA and AOB

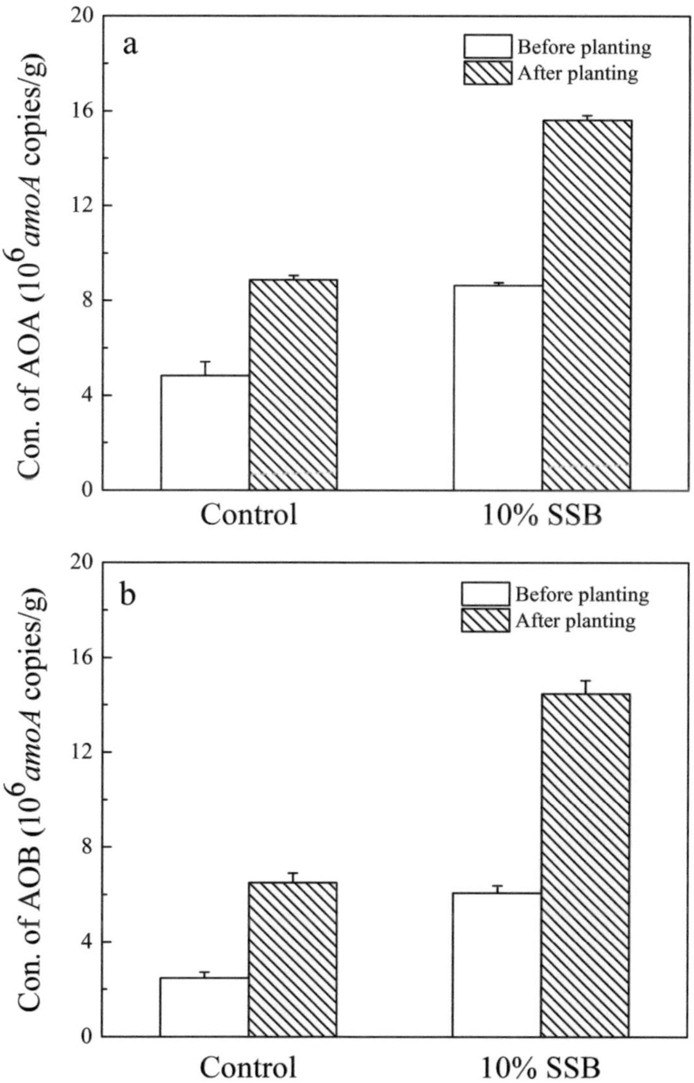

Figure 5.
Effects of SSB addition on the concentrations of AOA (a) and AOB (b) in soil.

concentrations in soil on adding 10% SSB reached 1.76 and 2.23 times as high as those of the control soil, respectively. These results show that SSB addition could effectively increase the concentrations of microorganisms associated with soil nitrification before and after planting.

To sum up, the influence of SSB on the microbiological property are as follows: on the one hand, SSB stored and supplied a large amount of nutrients by the bonding of nutrient cations and inorganic anions in soil with its surface functional groups; on the other hand, SSB changed the physiochemical property of soil and reduced the toxicity of contaminants to soil microorganisms [17].

3.4 Effects of adding SSB on Chinese cabbage growth

The weights of the aboveground and underground parts of Chinese cabbage are considered as important indicators that directly reflect the influence of the physical, chemical, and microbial properties of soil on plant growth. **Figure 6** shows the effects of adding SSB on the weight of Chinese cabbage. The weights of the aboveground and underground parts of Chinese cabbage increased with 10% SSB added to soil. The weight of edible aboveground part was 5.82 times and that of the underground part was 8.67 times as much as those from the control soil. These results can be explained by the fact that the addition of SSB brought the pH and EC of the original soil to suitable ranges for plant growth, and that the increases of the DHA activity, urease activity, bacteria concentration, and fungi concentration provided appropriate metabolic environment for soil microorganisms. This favorable metabolic environment further improved the microbial characteristics and forms a virtuous cycle [17]. In addition, SSB contains nutritive elements like K, P, and N at high concentrations, which increased the fertility of barren soil [9]. Therefore, the weights of Chinese cabbage increased significantly after SSB addition. This also showed that SSB had a positive effect on the growth of crop in barren soil.

3.5 Effects of SSB addition on HMs availability in Chinese cabbage and soil

Figure 7 shows the concentrations of HMs in the aboveground and underground parts of Chinese cabbage, respectively. For the aboveground part, the addition of SSB to soil significantly decreased the concentrations of Mn and Cd, and reduced the toxicity of Chinese cabbage in the edible part compared with the control group. For the underground part, the addition of SSB significantly decreased the

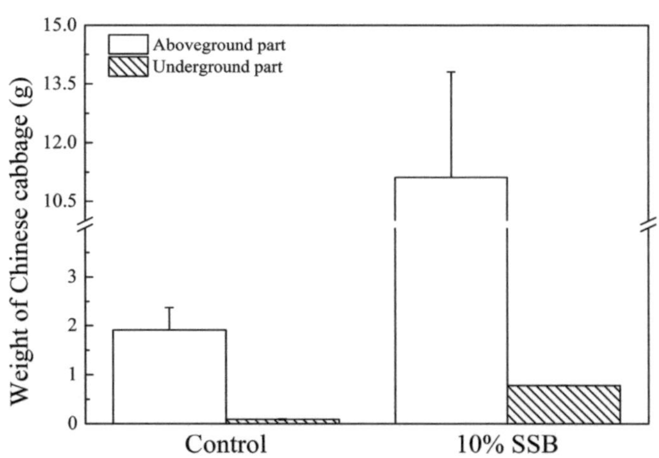

Figure 6.
Effects of adding SSB on the weight of Chinese cabbage planted.

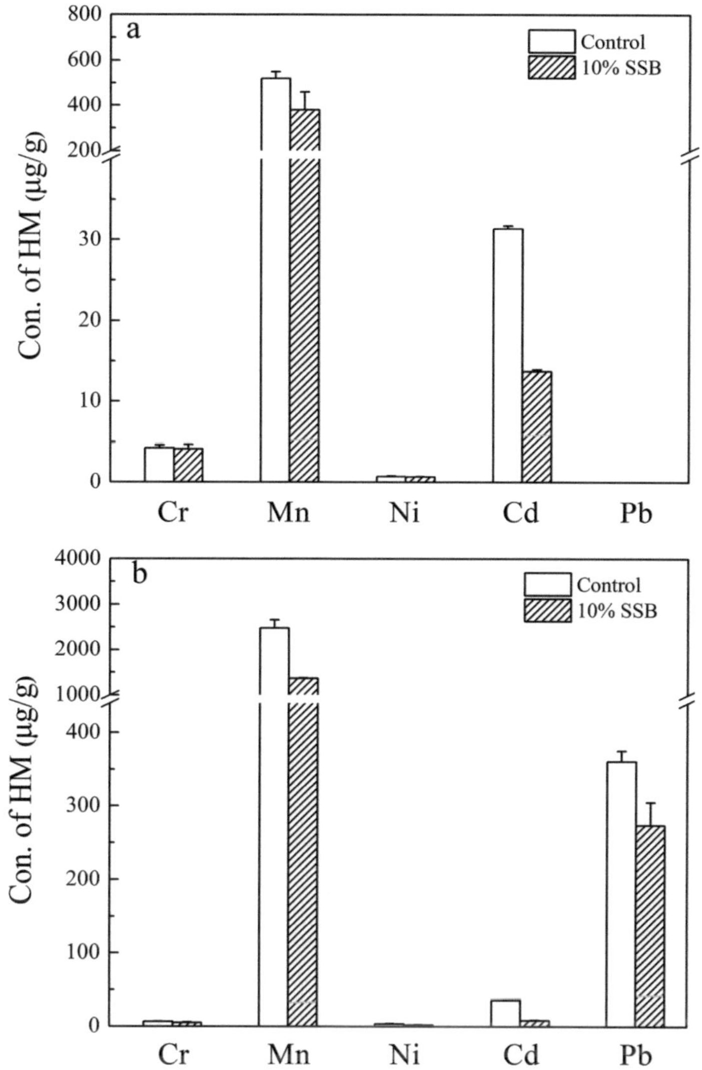

Figure 7.
Effects of adding SSB on the concentrations of HMs in aboveground part (a) and underground part (b) of Chinese cabbage planted.

concentrations of Mn, Pb, and Cd compared with the control group, which implied that the addition of SSB in soil inhibited the migration of HMs from soil to the underground part of Chinese cabbage.

It is widely accepted that the HMs in plant are entirely from the migration of the available HMs in the mixed soil during planting [4, 35]. Therefore, the concentrations of available HMs in soil before and after planting were measured to investigate the influence of SSB addition on the transfer of HMs, as shown in **Table 3**. The change rate of available HM concentration in soil after planting compared with that before planting was defined as:

$$\alpha = \frac{c_{s2} - c_{s1}}{c_{s1}} \times 100 \qquad (1)$$

where, α is the change rate of available HM concentration in soil after planting compared with that before planting, %; c_{s2} is the concentration of available HM in

HM	Condition	Before planting (μg/g)	After planting (μg/g)	α (%)
Cr	Control	8.24 ± 0.01	8.20 ± 0.38	−0.49
	10% SSB	6.80 ± 0.06	6.68 ± 0.29	−1.76
Mn	Control	0.35 ± 0.02	0.34 ± 0.02	−2.86
	10% SSB	0.34 ± 0.00	0.31 ± 0.01	−8.82
Ni	Control	7.59 ± 0.02	7.92 ± 0.03	+4.35
	10% SSB	7.10 ± 0.09	7.08 ± 0.68	−0.28
Cd	Control	4.88 ± 0.11	5.57 ± 0.22	+14.14
	10% SSB	4.62 ± 0.47	4.75 ± 0.15	+2.81
Pb	Control	0.36 ± 0.05	0.35 ± 0.06	−2.78
	10% SSB	0.27 ± 0.03	0.25 ± 0.02	−7.41

Table 3.
Concentrations and change rates of available HMs in soil before and after planting.

the soil after planting, μg/g; and c_{s1} is the concentration of available HM in the soil before planting, μg/g.

The addition of SSB decreased the concentrations of available Cr, Mn, Ni, Cd, and Pb in soil before planting, which is mostly because the fractions of HMs in SSB are more stable than those in soil. After planting, the concentrations of available Cr, Mn, and Pb in control soil decreased by 0.49, 2.86, and 2.78%, respectively, which indicated that these HMs were taken up by cabbages or migrated to more stable fractions during planting. Compared with the control soil, the addition of SSB reduced the transfer of the available HMs in soil during planting and the α value of Cr, Mn, Ni, Cd, and Pb decreased from −0.49, −2.86, 4.35, 14.14, and −2.78 to −1.76, −8.82, −0.28, 2.81, and −7.41%, respectively.

In order to investigate the effects of SSB addition on the migration of the available HMs in soil, the conversion rate of the content of available HM was defined as:

$$\eta = \frac{c_{s2} \cdot m_{s2} + c_{ca} \cdot m_{ca} + c_{cb} \cdot m_{cb} - c_{s1} \cdot m_{s1}}{c_{s1} \cdot m_{s1}} \times 100 \quad (2)$$

where η is the conversion rate of the content of available HM, %; c_{ca} and c_{cb} are the concentration of available HM in the aboveground and underground parts of Chinese cabbage after planting, respectively, μg/g; m_{s1} and m_{s2} are the mass of soil before and after planting, respectively, g; and m_{ca} and m_{cb} are the mass of the aboveground and underground parts of Chinese cabbage, respectively, g. When $\eta > 0$, the HM in soil transforms from the stable state to the available state after planting; and when $\eta < 0$, the HM in soil transforms from the available state to the stable state after planting. The conversion rates of the available HMs are shown in **Table 4**. The planting of Chinese cabbage in control soil promoted the immobilization of Cr and Pb and inhibited the immobilization of Mn, Ni, and Cd. Moreover, the addition of SSB increased the conversion rate of Mn compared with the control soil, which indicated that SSB addition could improve the migration of Mn to the available state. Mn plays an important role in the process of photosynthesis, respiration, protein synthesis, and hormone activation [36], which explains partly the effects of SSB addition on the weights of the aboveground and underground parts of Chinese cabbage. In addition, the conversion rates of Cr, Ni, and Cd after adding SSB decreased from −0.46, 4.35, and 14.40 to −1.62, −0.26, and 3.50% compared with the control soil, which indicated that the addition of SSB could promote the migration of Ni, Cd, and Cr from the available state to the stable state. In general, this

Condition	η (%)				
	Cr	Mn	Ni	Cd	Pb
Control	−0.46	66.65	4.35	14.40	−0.97
10% SSB	−1.62	302.60	−0.26	3.50	8.42

Table 4.
Conversion rates of the content of available HMs.

immobilization effect was closely related to the biochar properties and its effects on the microbial environment in soil. The surface of SSB has numerous rich alkaline groups such as alkyl negative ion, bicarbonate, carbonate, and phosphate [31], and its application in soil increased the pH, which led to the immobilization of the available HMs in soil [37, 38]. And, the interactions of SSB with the available HMs promoted the more stable transformation of HMs, and included the ion exchange between metal in soil and exchangeable metal in SSB, electrostatic attraction of anionic metal, electrostatic attraction of cation metal, and precipitation of metal [39]. Also, SSB has a good porous structure and can improve the microbial activity, which enhanced the transformation of microorganism on HMs [17]. Therefore, the addition of SSB could improve the immobilization of available HMs in soil.

4. Conclusions

SSB has better pH and EC, more developed pore structure, and higher concentrations of nutrient elements compared with the original soil. The addition of SSB could adjust the pH of mine soil from acidic to neutral and increase the EC of soil. Also, the addition of SSB increased the concentrations of enzyme and microorganisms. Therefore, the changes of the physiochemical property and microbial environment improved the growth of Chinese cabbage. The edible aboveground and the underground parts of cabbage in SSB-amended soil weighed 5.82 times and 8.67 times as much as those from the control group. Moreover, the addition of SSB promoted the migration of Cr, Ni, and Cd from the available state to the more stable state due to the special properties of SSB and changes of soil environment. To sum up, SSB has positive effects on the planting in barren soil.

Acknowledgements

The authors would like to thank Xiang Zhang for his valuable help. This work was supported by financial support received from the Industry Leading Key Projects of Fujian Province (2015H0044), the China-Japanese Research Cooperative Program (2016YFE0118000), the Scientific and Technological Major Special Project of Tianjin City (16YFXTSF00420), and the Key Project of Young Talent of IUE, CAS (IUEZD201402).

Author details

Guangwei Yu[1*], Shengyu Xie[1,2], Jianli Ma[3], Xiaofu Shang[4], Yin Wang[1], Cheng Yu[5], Futian You[1], Xiaoda Tang[1], Héctor U. Levatti[6], Lanjia Pan[1,2], Jie Li[1,2] and Chunxing Li[1]

1 CAS Key Laboratory of Urban Pollutant Conversion, Institute of Urban Environment, Chinese Academy of Sciences, Xiamen, China

2 University of Chinese Academy of Sciences, Beijing, China

3 Tianjin Academy of Environmental Sciences, Tianjin, China

4 Tianjin Huankelijia Environment Remediation Technology Co., Ltd., Tianjin, China

5 Fujian Academy of Building Research, Fuzhou, China

6 College of Engineering, Swansea University, Swansea, China

*Address all correspondence to: gwyu@iue.ac.cn

References

[1] Zhang Q, Hu J, Lee DJ, et al. Sludge treatment: Current research trends. Bioresource Technology. 2017;**243**:1179-1172. DOI: 10.1016/j.biortech.2017.07.070

[2] Shao J, Yuan X, Leng L, et al. The comparison of the migration and transformation behavior of heavy metals during pyrolysis and liquefaction of municipal sewage sludge, paper mill sludge, and slaughterhouse sludge. Bioresource Technology. 2015;**198**:16-22. DOI: 10.1016/j.biortech.2015.08.147

[3] Chen H, Yan SH, Ye ZL, et al. Utilization of urban sewage sludge: Chinese perspectives. Environmental Science and Pollution Research. 2012;**19**(5):1454-1463. DOI: 10.1007/s11356-012-0760-0

[4] Khan S, Chao C, Waqas M, et al. Sewage sludge biochar influence upon rice (*Oryza sativa* L) yield, metal bioaccumulation and greenhouse gas emissions from acidic paddy soil. Environmental Science and Technology. 2013;**47**(15):8624-8632. DOI: 10.1021/es400554x

[5] Chen D, Yin L, Wang H, et al. Pyrolysis technologies for municipal solid waste: A review. Waste Management. 2014;**34**(12):2466-2486. DOI: 10.1016/j.wasman.2014.08.004

[6] Ding HS, Jiang H. Self-heating co-pyrolysis of excessive activated sludge with waste biomass: Energy balance and sludge reduction. Bioresource Technology. 2013;**133**:16-22. DOI: 10.1016/j.biortech.2013.01.090

[7] Atkinson CJ, Fitzgerald JD, Hipps NA. Potential mechanisms for achieving agricultural benefits from biochar application to temperate soils: A review. Plant and Soil. 2010;**337**(1):1-18. DOI: 10.1007/s11104-010-0464-5

[8] Zhu Q, Wu J, Wang L, et al. Effect of biochar on heavy metal speciation of paddy soil. Water, Air, and Soil Pollution. 2015;**226**(12):429. DOI: 10.1007/s11270-015-2680-3

[9] Julie M, Marco R, Diego M, et al. Maize yield and nutrition during 4 years after biochar application to a Colombian savanna oxisol. Plant and Soil. 2010;**333**(1-2):117-128. DOI: 10.1007/s11104-010-0327-0

[10] Song XD, Xue XY, Chen DZ, et al. Application of biochar from sewage sludge to plant cultivation: Influence of pyrolysis temperature and biochar-to-soil ratio on yield and heavy metal accumulation. Chemosphere. 2014;**109**:213-220. DOI: 10.1016/j.chemosphere.2014.01.070

[11] Zhang X, Wang H, He L, et al. Using biochar for remediation of soils contaminated with heavy metals and organic pollutants. Environmental Science and Pollution Research. 2013;**20**(12):8472-8483. DOI: 10.1007/s11356-013-1659-0

[12] Méndez A, Gómez A, Paz-Ferreiro J, et al. Effects of sewage sludge biochar on plant metal availability after application to a Mediterranean soil. Chemosphere. 2012;**89**(11):1354-1359. DOI: 10.1016/j.chemosphere.2012.05.092

[13] Hossain MK, Strezov V, Yin Chan K, et al. Agronomic properties of wastewater sludge biochar and bioavailability of metals in production of cherry tomato (*Lycopersicon esculentum*). Chemosphere. 2010;**78**(9):1167-1171. DOI: 10.1016/j.chemosphere.2010.01.009

[14] Li J, Pan L, Yu G, et al. Preparation of adsorption ceramsite derived from sludge biochar. Environmental Science. 2017;**38**(9):3970-3978. DOI: 10.13227/j.hjkx.201702161

[15] Li J, Yu GW, Xie SY, et al. Immobilization of heavy metals in ceramsite produced from sewage sludge biochar. Science of the Total Environment. 2018;**628-629**:131-140. DOI: 10.1016/j.scitotenv.2018.02.036

[16] Wang X, Li C, Zhang B, et al. Migration and risk assessment of heavy metals in sewage sludge during hydrothermal treatment combined with pyrolysis. Bioresource Technology. 2016;**221**:560-567. DOI: 10.1016/j.biortech.2016.09.069

[17] Zhu XM, Chen BL, Zhu LZ, et al. Effects and mechanisms of biochar-microbe interactions in soil improvement and pollution remediation: A review. Environmental Pollution. 2017;**227**:98-115. DOI: 10.1016/j.envpol.2017.04.032

[18] Pignataro A, Moscatelli MC, Mocali S, et al. Assessment of soil microbial functional diversity in a coppiced forest system. Applied Soil Ecology. 2012;**62**:115-123. DOI: 10.1016/j.apsoil.2012.07.007

[19] Li ZW, Wang XD, Lin JJ, et al. Transformation of nitrogen, phosphorus, potassium and heavy metals during sewage sludge biochar preparation. Chinese Journal of Environmental Engineering. 2016;**10**(3):1392-1399

[20] Lindsay WL, Norvell WA. Development of a DTPA soil test for zinc, iron, manganese, and copper. Soil Science Society of America Journal. 1978;**42**(3):421-428. DOI: 10.2136/sssaj 1978.03615995004200030009x

[21] Singh J, Singh DK. Dehydrogenase and phosphomonoesterase activities in groundnut (*Arachis hypogaea* L.) field after diazinon, imidacloprid and lindane treatments. Chemosphere. 2005;**60**(1):32-42. DOI: 10.1016/j.chemosphere.2004.11.096

[22] Sudkolai ST, Nourbakhsh F. Urease activity as an index for assessing the maturity of cow manure and wheat residue vermicomposts. Waste Management. 2017;**64**:63-66. DOI: 10.1016/j.wasman.2017.03.011

[23] Atkins SD, Clark IM, Pande S, et al. The use of real-time PCR and species-specific primers for the identification and monitoring of Paecilomyces lilacinus. FEMS Microbiol Ecology. 2005;**51**(2):257-264. DOI: 10.1016/j.femsec.2004.09.002

[24] Hilioti Z, Michailof CM, Valasiadis D, et al. Characterization of castor plant-derived biochars and their effects as soil amendments on seedlings. Biomass Bioenergy. 2017;**105**:96-106. DOI: 10.1016/j.biombioe.2017.06.022

[25] Huang HJ, Yang T, Lai FY, et al. Co-pyrolysis of sewage sludge and sawdust/rice straw for the production of biochar. Journal of Analytical and Applied Pyrolysis. 2017;**125**:61-68. DOI: 10.1016/j.jaap.2017.04.018

[26] Li J, Yu G, Pan L, et al. Study of ciprofloxacin removal by biochar obtained from used tea leaves. Journal of Environmental Sciences. 2018;**73**:20-30. DOI:10.1016/j.jes.2017.12.024

[27] Qadeer S, Anjum M, Khalid A, et al. A dialogue on perspectives of biochar applications and its environmental risks. Water, Air, and Soil Pollution. 2017;**228**(8):281. DOI: 10.1007/s11270-017-3428-z

[28] Adeleke R, Nwangburuka C, Oboirien B. Origins, roles and fate of organic acids in soils: A review. South African Journal of Botany. 2017;**108**:393-406. DOI: 10.1016/j.sajb.2016.09.002

[29] Belyaeva ON, Haynes RJ, Sturm EC. Chemical, physical and microbial properties and microbial diversity

in manufactured soils produced from co-composting green waste and biosolids. Waste Management. 2012;**32**(12):2248-2257. DOI: 10.1016/j.wasman.2012.05.034

[30] Corwin DL, Lesch SM. Application of soil electrical conductivity to precision agriculture. Agronomy Journal. 2003;**95**:455-471. DOI: 10.2134/agronj2003.0455

[31] Yuan JH, Xu RK, Zhang H. The forms of alkalis in the biochar produced from crop residues at different temperatures. Bioresource Technology. 2011;**102**(3):3488-3497. DOI: 10.1016/j.biortech.2010.11.018

[32] Chendrayan K, Adhya TK, Sethunathan N. Dehydrogenase and invertase activities of flooded soils. Soil Biol and Biochemistry. 1980;**12**(3):271-273. DOI: 10.1016/0038-0717(80)90073-5

[33] Guettes R, Dott W, Eisentraeger A. Determination of urease activity in soils by carbon dioxide release for ecotoxicological evaluation of contaminated soils. Ecotoxicology. 2002;**11**(5):357-364. DOI: 10.1023/A:1020509422554

[34] Luo J, Lin WT. Research progress of ammonia-oxidizing archaea. Journal of South China University of Technology. 2013;**41**(12):107-114. DOI: 10.3969/j.issn.1000-565X.2013.12.018

[35] Zeng F, Ali S, Zhang H, et al. The influence of pH and organic matter content in paddy soil on heavy metal availability and their uptake by rice plants. Environmental Pollution. 2011;**159**(1):84-91. DOI: 10.1016/j.envpol.2010.09.019

[36] Shao JF, Yamaji N, Shen R F, et al. The key to Mn homeostasis in plants: Regulation of Mn transporters. Trends in Plant Science. 2017;**22**(3):215-224. DOI: 10.1016/j.tplants.2016.12.005

[37] Kołodyńska D, Wnętrzak R, Leahy JJ, et al. Kinetic and adsorptive characterization of biochar in metal ions removal. Chemical Engineering Journal. 2012;**197**:295-305. DOI: 10.1016/j.cej.2012.05.025

[38] Almaroai YA, Usman ARA, Ahmad M, et al. Effects of biochar, cow bone, and eggshell on Pb availability to maize in contaminated soil irrigated with saline water. Environmental Earth Sciences. 2014;**71**(3):1289-1296. DOI: 10.1007/s12665-013-2533-6

[39] Ahmad M, Rajapaksha A U, Lim JE, et al. Biochar as a sorbent for contaminant management in soil and water: A review. Chemosphere. 2014;**99**:19-33. DOI: 10.1016/j.chemosphere.2013.10.071

Section 2

Bioremediation with Biochar

Chapter 5

Application of Biochar to Enzyme Carrier for Stress Tolerance of Enzymes

Hidetaka Noritomi

Abstract

Biochar showed the high affinity to enzymes, and enzymes were sufficiently adsorbed on the surface of biochar. Enzymes were highly stabilized in water at high temperatures by adsorbing enzymes on biochar. The remaining activity of lysozyme adsorbed on adzuki bean charcoal showed around 50% after heat treatment at 90°C for 30 min though that of free lysozyme was almost lost. Likewise, the stability of enzymes was enhanced in organic solvents by adsorbing enzymes on biochar. The conformation of α-chymotrypsin adsorbed on bamboo charcoal was hardly influenced by organic solvents, while that of free α-chymotrypsin was strongly dependent of the kind of organic solvents. Moreover, the adsorption of α-chymotrypsin on bamboo charcoal improved the transesterification of *N*-acetyl-L-tyrosine ethyl ester with *n*-butanol in organic solvents. The transesterification rate of α-chymotrypsin adsorbed on bamboo charcoal was about 760 times higher than that of free α-chymotrypsin in *n*-butyl acetate.

Keywords: biochar, enzyme, enzyme carrier, adsorption, stress tolerance

1. Introduction

An enormous amount of greenhouse gas such as CO_2 has recently been emitted from industries and thereby has caused serious global warming problems [1, 2]. Accordingly, the application of biomass materials, which are carbon neutral, to energies and functional materials, is crucial to reduce greenhouse gas emissions [3, 4]. However, most of biomass materials such as forestry residues have hardly been utilized in the field of functional materials. Accordingly, the development in the high value-added application of biomass materials has been desired to provide the multiple effective utilization system of biomass materials.

Enzymes are biocatalysts, which exhibit their outstanding biological activity under mild conditions, and have widely been used in pharmacy, biotechnology, and chemical industry [5–7]. Typical applications of enzymes are biotransformation, biosensor, biofuel cell, and so on. Enzymes are generally stable in a cell. However, they are gradually denatured and inactivated under various physical and chemical stresses such as heat, organic solvents, and so on [8]. In order to enhance the stability of enzymes used in vitro, enzyme immobilization, where enzyme molecules are attached to solid carriers, has widely been used [9–11]. The main required features of enzyme carriers are chemical stability, thermal stability, insolubility under reaction conditions, high affinity to enzymes, biocompatibility, the presence of reactive functional groups, availability, low price, regeneration, reusability, and so on. When enzymes are immobilized on carriers through adsorption, the catalytic activity, specificity, and stability of enzymes are influenced by the nature of carriers. Accordingly, the performance of enzymes can be enhanced by selecting an appropriate carrier.

A large quantity of the world's oldest biochar was excavated from the cave of Kara Iwatani of Hijikawacho, Ozu-shi, Ehime, Japan, with a beast bone and the human bone piece 300,000 years ago. Since the ancient period, the biochar has been used not only as a fuel but also as a soil conditioner to support human life for a long time in the world [12]. Consequently, as the biocompatibility of biochar can be expected, we have examined the application of biochar to enzyme carriers. As a result, we have found that enzymes are effectively adsorbed on biochar [13, 14], and biochar-adsorbed enzymes exhibit the high thermal stability in water [15–18]. Moreover, we have reported that the adsorption of enzymes on biochar sufficiently improves the enzyme activity in organic solvents [19–21].

In the chapter, the characterization of biochar, the adsorption of enzyme on biochar, the high temperature-tolerant property of biochar-adsorbed enzymes, and the organic solvent-tolerant property of biochar-adsorbed enzymes are discussed.

2. Heat tolerance of biochar-adsorbed enzyme

2.1 Preparation and characterization of biochar

Biochar has been prepared by pyrolyzing plant biomass waste such as bamboo waste at low temperatures under nitrogen atmosphere to produce functional groups, which were used as a binding site for the adsorption of enzymes (**Figure 1**) [13, 14]. The emission of carbon dioxide was reduced since plant biomass waste was not burned through the preparation of biochar. Moreover, the energy cost of the present preparation of biochar was suppressed, compared to that of the conventional preparation of charcoal. Consequently, the present preparation of biochar was a low-cost and environmentally benign process.

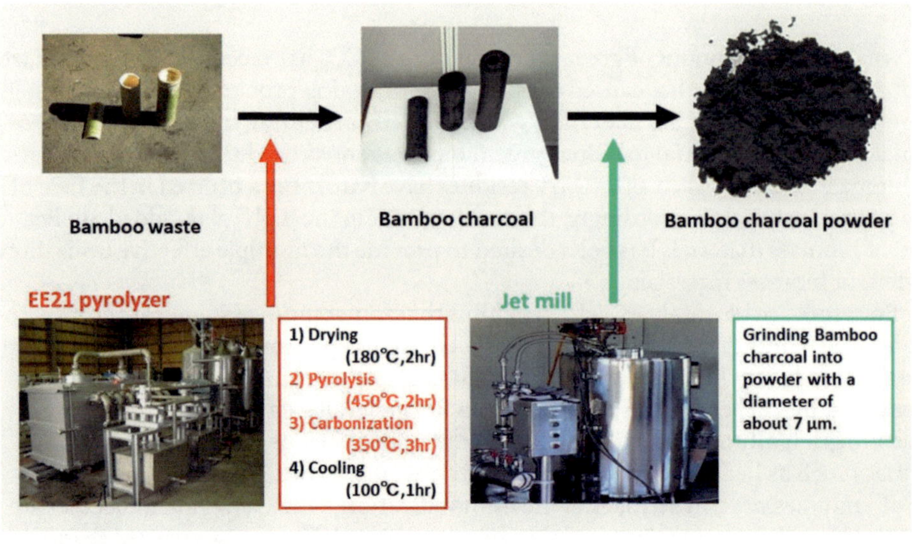

Figure 1.
Preparation process of biochar.

In order to observe the surface of biochar, we have examined SEM images [14]. As seen in **Figure 2**, the morphology of biochar was strongly dependent upon the kind of raw materials. The roughness of biochar was remarkably low, and any pores were not observed at the magnification measured in the present work. The surface of bamboo charcoal was smoother than that of any other charcoal. Moreover, Raman spectra of biochar showed that the structure of biochar was amorphous [13].

Table 1 shows the textural parameters of biochar obtained from low-temperature (−196°C) nitrogen adsorption isotherms, which allow the calculation of specific surface area, specific pore volume, and pore diameter peak [14]. In the table, the specific area of adzuki bean charcoal depicted the value obtained from the CO_2 isotherm. The specific surface area and specific pore volume of biochar showed much small, compared with that of conventional activated carbon. The carbonizing temperature affects the surface property of charcoal [22]. The specific pore volume increases with an increase in carbonizing temperature. Consequently, the pore of biochar was not formed enough at low temperatures.

Figure 3 shows CP/MAS ^{13}C-NMR spectra of biochar [13]. Aromatic carbon (140–141, 131 ppm) was mainly detected, and C=O (200 ppm), COOH, CHO (175–190 ppm), and aromatic oxygen (150–153, 145–146 ppm) were also detected. Moreover, in order to assess the chemical property of the surface of biochar, the measurement on X-ray photoelectron spectroscopy (XPS) was carried out [19]. **Figure 4** shows the elemental ratio of the surface of biochar detected by XPS. The main element was carbon atom, and oxygen and nitrogen atoms also existed on the surface of biochar to some extent. The ratios of oxygen and nitrogen atoms in adzuki bean charcoal were greater than those in any other charcoal. Narrow scan spectra of XPS showed C—C, C—H, C—O, O—C—O, C=O, COOH, and C—N, as seen in **Figure 5**. Many radical species due to functional groups containing oxygen atoms, which are formed by thermal decomposition of cellulose and hemicelluloses, are detected in charcoals carbonized at 500°C by the measurement of electron spin resonance, and functional groups decrease with increasing carbonization temperature [22, 23].

Figure 6 shows the relationship of the ζ potential of biochar with the solution pH [15]. The ζ potential of adzuki bean charcoal drastically decreased with

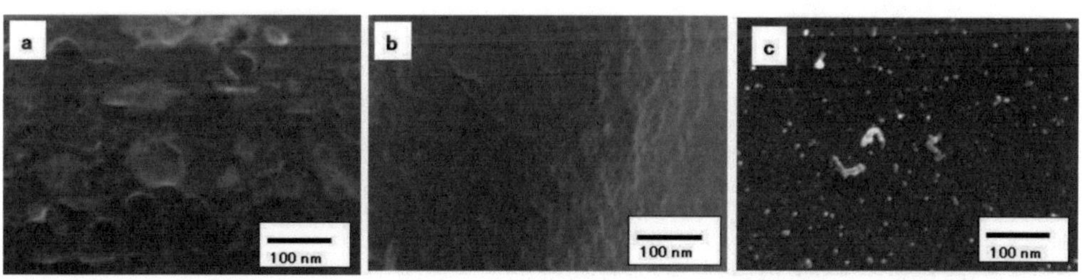

Figure 2.
SEM images of (a) adzuki bean charcoal, (b) bamboo charcoal, and (c) wood charcoal.

Biochar	Specific surface area [m²/g]	Pore volume [cm³/g]	Pore diameter peak [nm]
Adzuki bean charcoal	204[a]	—	—
Bamboo charcoal	294	0.041	Less than 2.6
Wood charcoal	117	0.025	Less than 2.6

[a]*Specific area of adzuki bean charcoal was obtained from the CO_2 isotherm.*

Table 1.
Structural characteristics of biochar.

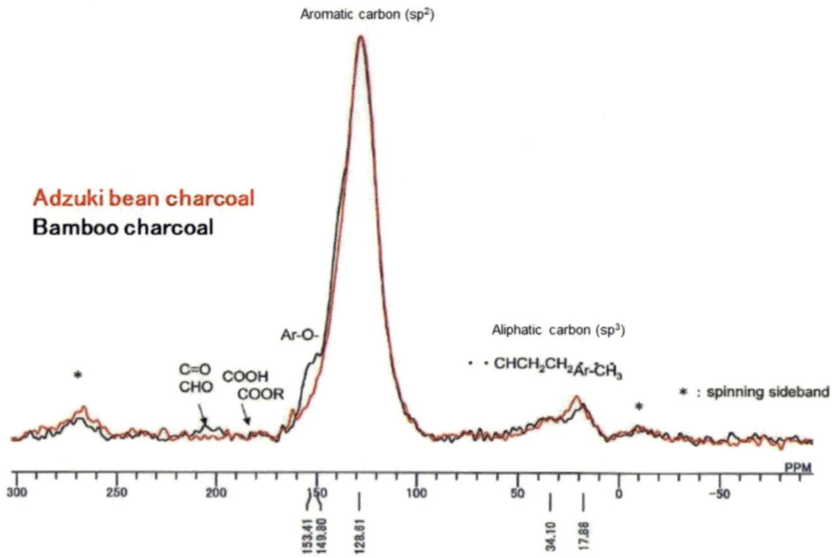

Figure 3.
CP/MAS ^{13}C-NMR spectra of biochar.

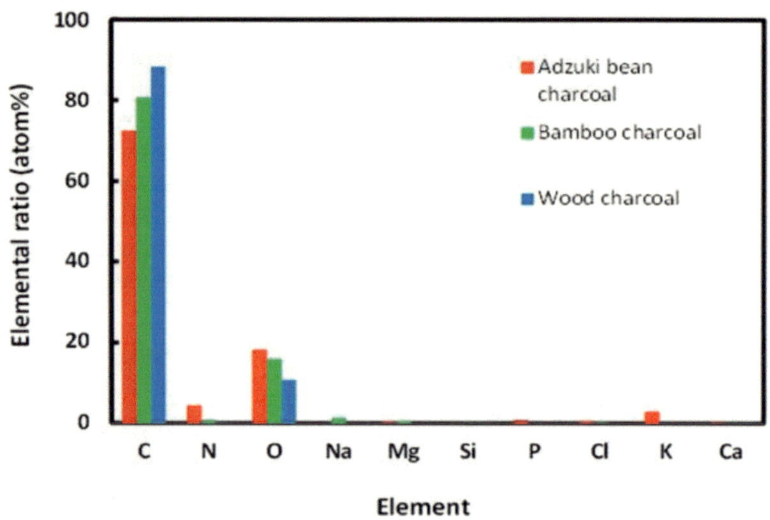

Figure 4.
The elemental ratio of the surface of biochar detected by XPS.

increasing the pH value, exhibiting a negative value above pH 4, drops till pH 7, and was almost constant in the alkaline region. The pH dependence of the ζ potential of bamboo charcoal exhibited the same tendency to the case of adzuki bean charcoal.

2.2 Adsorption of enzymes on biochar

Figure 7 shows the time course of the amount of lysozyme adsorbed on adzuki bean charcoal at pH 7 and 25°C when hen egg white lysozyme was employed as a model enzyme [16]. The amount of lysozyme adsorbed on adzuki bean charcoal increased with an increase in adsorption time, reached a plateau around 24 h, and

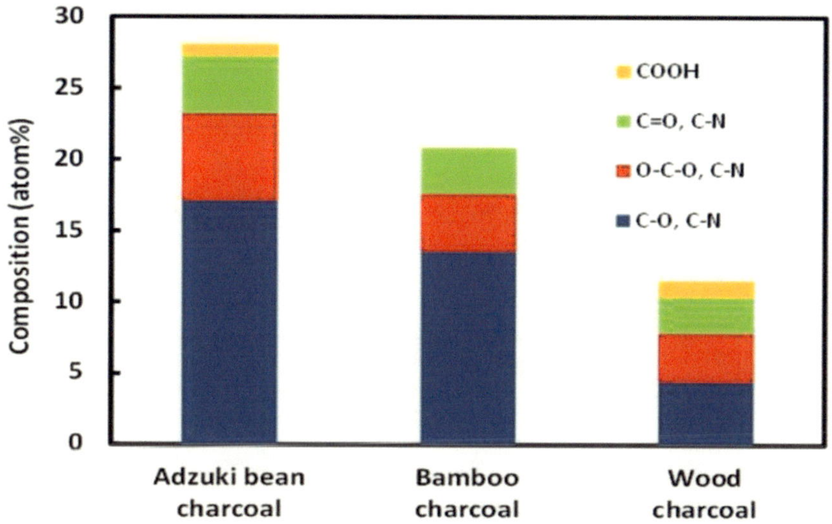

Figure 5.
The chemical bond ratio of biochar obtained from narrow scan spectra of XPS.

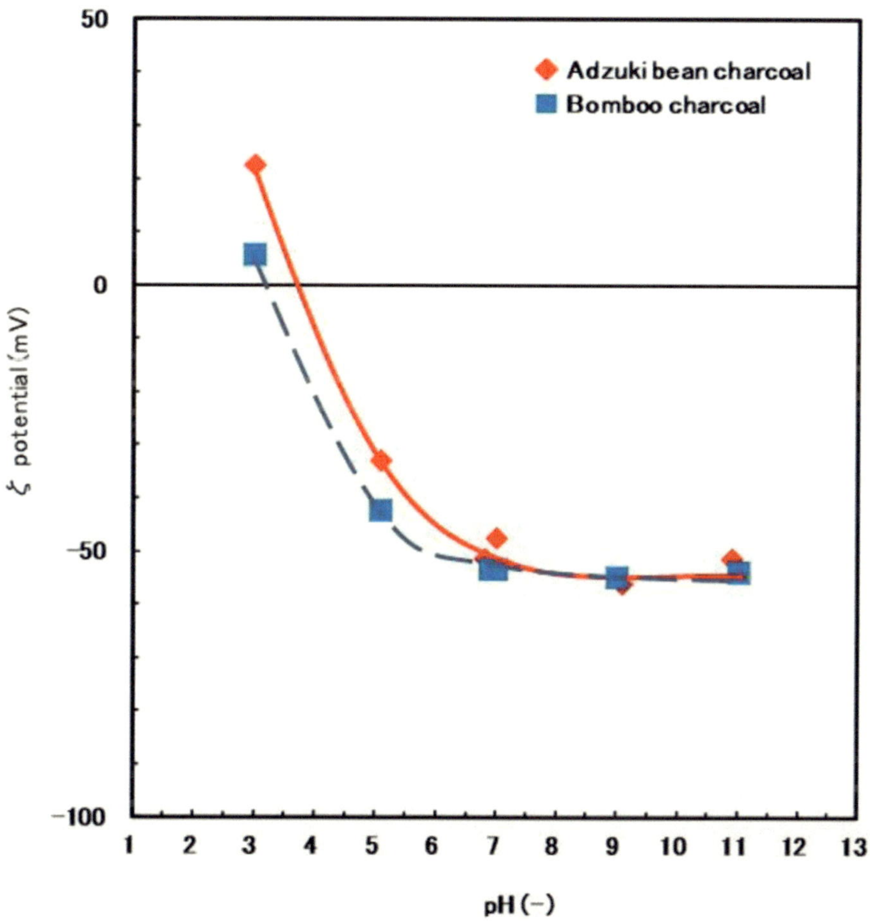

Figure 6.
Effect of solution pH on ζ potential of biochar.

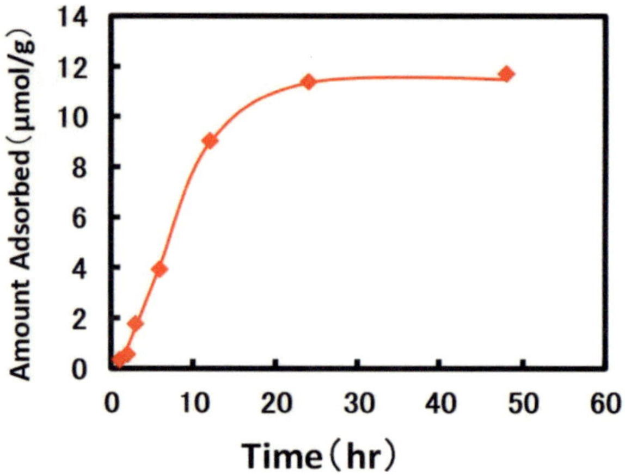

Figure 7.
Time dependence of amount of lysozyme adsorbed onto adzuki bean charcoal.

Biochar	Amount of enzymes adsorbed (μmol/g)	
	Lysozyme	α-Chymotrypsin
Adzuki bean charcoal	11	17
Bamboo charcoal	9	9.8
Wood charcoal	12	21

Table 2.
Amount of enzymes adsorbed on biochar.

was 11 μmol/g (0.16 g/g). As overall concentration of adzuki bean charcoal was 3 g/L in an aqueous solution, overall lysozyme concentration in the aqueous solution corresponded to 33 μM (0.48 mg/mL). From this result, the adsorption of lysozyme on adzuki bean charcoal was carried out for 24 h.

As seen in **Table 2**, the amount of enzymes adsorbed on biochar was strongly dependent on the kinds of enzymes and/or biochars [14, 15]. The amount of lysozyme adsorbed on biochar was almost the same among three different charcoals although the specific surface area of adzuki bean charcoal or bamboo charcoal was more than twice larger than wood charcoal, as seen in **Table 1**.

Figure 8 shows the adsorption isotherms of lysozyme on biochar. These isotherms gradually increased [13]. The amount adsorbed on adzuki bean charcoal exhibited large, compared to the amount adsorbed on bamboo charcoal. The curves in the figure were the fitting lines with Freundlich adsorption isotherm equation (Eq. 1).

$$W = K_F C^{1/n} \tag{1}$$

Here, W and C are the amount adsorbed and the concentration of lysozyme, respectively. K_F and n are experimental constants [24]. The curves of adzuki bean charcoal and bamboo charcoal had the correlation constants (r^2) of 0.974 and 0.998, respectively. On the other hand, the data were fitted by Langmuir adsorption isotherm equation; the curves of adzuki bean charcoal and bamboo charcoal depicted the correlation constants (r^2) of 0.721 and 0.694, respectively.

Figure 9 shows the relationship of the amount of lysozyme adsorbed on biochar with the pH value of aqueous solutions at 25°C [14]. The curve of the amount

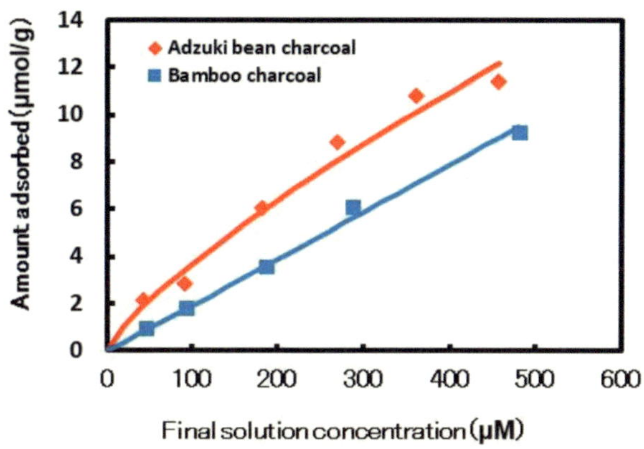

Figure 8.
Adsorption isotherms of lysozyme onto biochar; adsorption was carried out by incubating buffer solution (pH 7) containing a certain amount of lysozyme and 3 g/L biochar at 120 rpm and 25°C for 24 h.

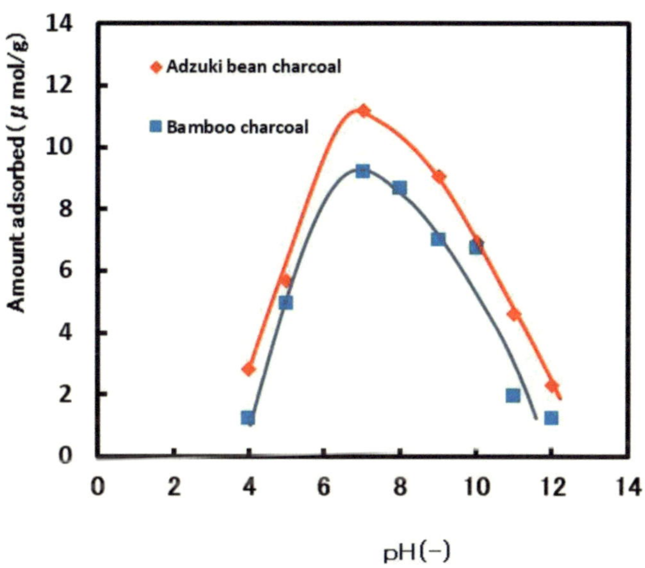

Figure 9.
Effect of pH on the amount of lysozyme adsorbed onto biochar; adsorption was carried out by incubating buffer solution (appropriate pH) containing 500 μM lysozyme and 3 g/L biochar at 120 rpm and 25°C for 24 h.

adsorbed on adzuki bean charcoal had the optimum value around neutral pH, similar to the case of bamboo charcoal. The net charge of protein molecules alters with the pH of aqueous solutions. Since the isoelectric point (pI) of lysozyme is 11, the net charge of lysozyme becomes more positive below pH 11. Concerning the ζ potential of biochar, the ζ potential of adzuki bean charcoal drastically decreased with increasing the pH value, exhibited a negative value above pH 4, dropped till pH 7, and was almost constant in the alkaline region, as shown in **Figure 6**. The pH dependence of the ζ potential of bamboo charcoal exhibited the same tendency to the case of adzuki bean charcoal. When the pH value was around the pI of lysozyme or the pH where the ζ potential of biochar approached 0 volts, a dramatic decrease in the amount of lysozyme adsorbed on biochar was observed. On the other hand, in the vicinity of neutral pH where lysozyme and the surface of biochar were charged positively and negatively, respectively, the high amount of adsorption tended to be obtained. Consequently, these results indicate that the electrostatic interaction between the positively charged lysozyme and the negatively charged surface of biochar mainly contributes to the adsorption.

2.3 Heat stress tolerance of enzymes adsorbed on biochar

Modest heating causes enzymes dissolved in an aqueous solution to be denatured and inactivated by unfolding of enzyme molecules due to the disruption of weak interactions such as ionic bonds, hydrogen bonds, and hydrophobic interactions, which are prime determinants of enzyme tertiary structures [25]. In order to assess the heat stress tolerance of enzymes adsorbed on biochar, an aqueous solution containing lysozyme adsorbed on adzuki bean charcoal was incubated at high temperatures [15]. **Figure 10** shows photographs of aqueous solutions containing free lysozyme, the mixture of lysozyme and adzuki bean charcoal, and lysozyme adsorbed on adzuki bean charcoal before and after heat treatment was carried out at 90°C for 30 min, while under such heat conditions, raw eggs become hard-boiled eggs. The solution of free lysozyme immediately became turbid since thermally denatured enzymes were drastically aggregated by heat, as shown in **Figure 10(d)**. The enzyme aggregation is precipitated above 10 μM lysozyme [26]. The formation of enzyme aggregation was enhanced since the present concentration of lysozyme was 33 μM. Adzuki bean charcoal was easily dispersed in an aqueous solution due to the good wettability to water as seen in **Figure 10(b)**. Likewise, the mixture of lysozyme and adzuki bean charcoal was immediately precipitated by heat treatment due to the aggregation of denatured enzymes, as shown in **Figure 10(e)**. On the other hand, lysozyme adsorbed on adzuki bean charcoal was easily dispersed in an aqueous solution, as seen in **Figure 10(c)**. After the heat treatment, lysozyme adsorbed on adzuki bean charcoal was sufficiently dispersed in the solution, and the white enzyme aggregation and the cohesion among adzuki bean charcoals adsorbing lysozyme were not observed in the solution, as shown in **Figure 10(f)**. When enzymes dissolved in an aqueous solution are placed at high temperatures, most of enzymes are instantaneously unfolded by the disruption of weak interactions consisting of ionic bonds, hydrogen bonds, and hydrophobic interactions of enzymes [25, 27]. Additionally, unfolded enzymes are aggregated with each other, and the chemical deterioration reactions occur in unfolded enzymes. In particular, enzyme aggregation easily occurs upon the exposure of the hydrophobic surfaces of an enzyme, and this phenomenon becomes the major problem because of the fast

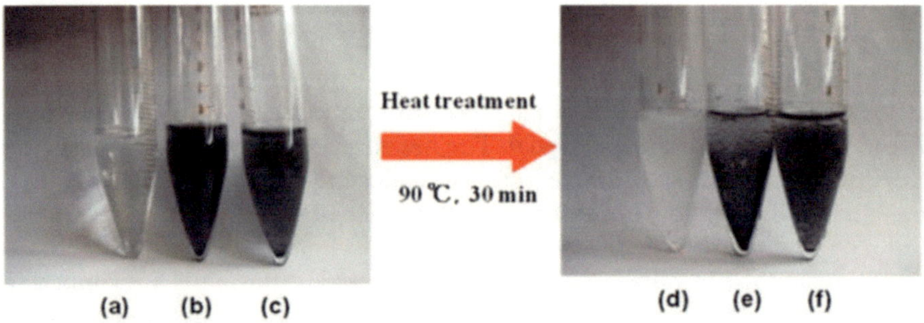

Figure 10.
Photographs of lysozyme solutions before and after heat treatment at 90°C for 30 min: (a) an aqueous solution containing free lysozyme before heat treatment, (b) an aqueous solution containing free lysozyme solution and adzuki bean charcoal before heat treatment, (c) an aqueous solution containing lysozyme adsorbed on adzuki bean charcoal before heat treatment, (d) an aqueous solution containing free lysozyme after heat treatment, (e) an aqueous solution containing free lysozyme solution and adzuki bean charcoal after heat treatment, and (f) an aqueous solution containing lysozyme adsorbed on adzuki bean charcoal after heat treatment.

irreversible inactivation. The adsorption of lysozyme on adzuki bean charcoal could inhibit the formation of enzyme aggregation.

When the remaining activity is defined as the ratio of the activity of lysozyme after heat treatment to that before heat treatment, time courses of remaining activities of free lysozyme and lysozyme adsorbed on adzuki bean charcoal through the heat treatment at pH 7.0 and 90°C are shown in **Figure 11** [16]. The remaining activities of free lysozyme and lysozyme adsorbed on adzuki bean charcoal decreased with an increase in time. As shown in **Figure 11**, the remaining activities of free and adsorbed lysozyme exhibited the correlation of first-order kinetics with heat treatment time. **Table 3** shows inactivation rate constants and half-lives of free and adsorbed lysozymes obtained from the curve fitting in **Figure 11**. The half-life of adsorbed lysozyme was seven times greater than that of free lysozyme. The remaining activity of free lysozyme was almost lost after heat treatment for 30 min, and the remaining activity in the mixture of lysozyme and adzuki bean charcoal exhibited 2%, while the remaining activity of lysozyme adsorbed on adzuki bean charcoal showed around 50%. The robust thermal stability of adsorbed lysozyme may be attributable to the suitable interaction of lysozyme with the surface of adzuki bean charcoal.

To extend our study, the remaining activities of lysozyme adsorbed on biochar obtained from several kinds of plant biomass wastes have been measured after heat treatment at 90°C for 30 min. Lysozyme adsorbed on bamboo charcoal or wood charcoal exhibited the high thermal stability, similar to the case of lysozyme-adsorbed adzuki bean charcoal, as shown in **Figure 12**. On the other hand, the

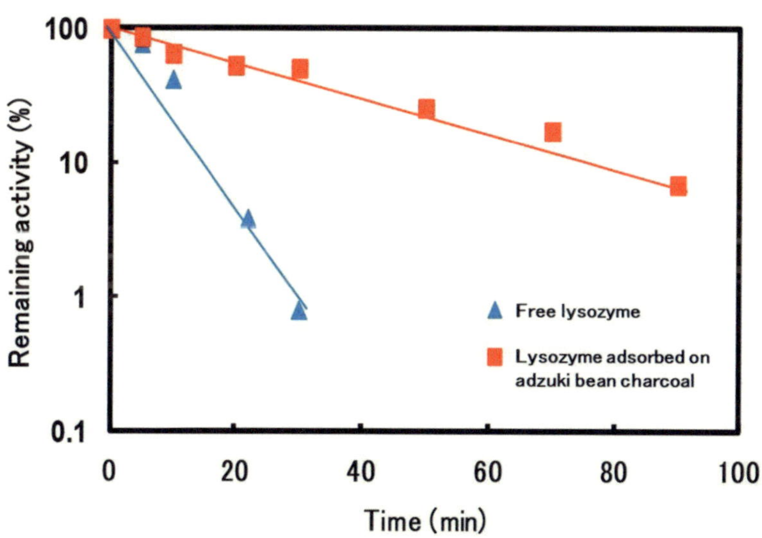

Figure 11.
Time course of remaining activity of free lysozyme and lysozyme adsorbed on adzuki bean charcoal through the heat treatment at pH 7.0 and 90°C.

Samples	Rate constant (min^{-1})	Half life (min)
Free lysozyme	0.168	4
Lysozyme adsorbed on adzuki bean charcoal	0.027	28

Table 3.
Rate constants and half lives of inactivation of lysozyme at 90°C.

mixture of lysozyme and bamboo charcoal or wood charcoal showed several percent of remaining activity.

The solution pH generally affects the activity and stability of enzymes in aqueous solutions [5]. **Figure 13** shows the remaining activity of lysozyme adsorbed on bamboo charcoal against the solution pH of adsorption medium. The remaining activity of bamboo charcoal-adsorbed lysozyme was markedly influenced by the pH of adsorption medium and showed the maximum value at pH 5.

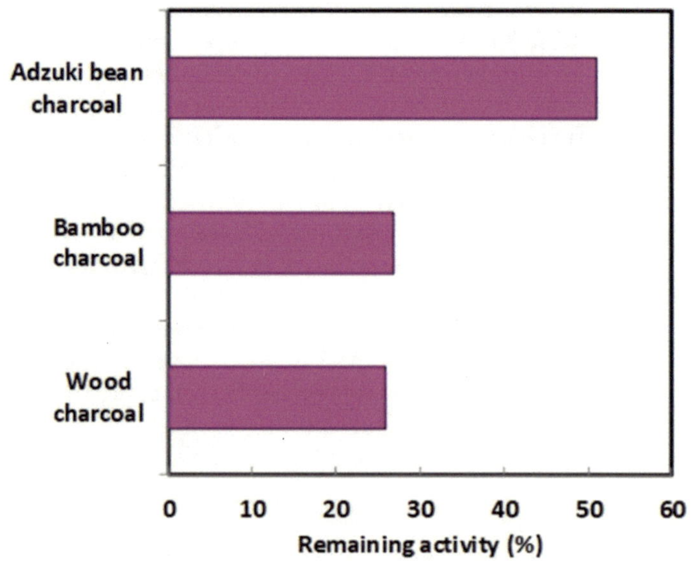

Figure 12.
Effect of kind of biochar on remaining activity of lysozyme adsorbed on biochar after heat treatment at 90°C for 30 min.

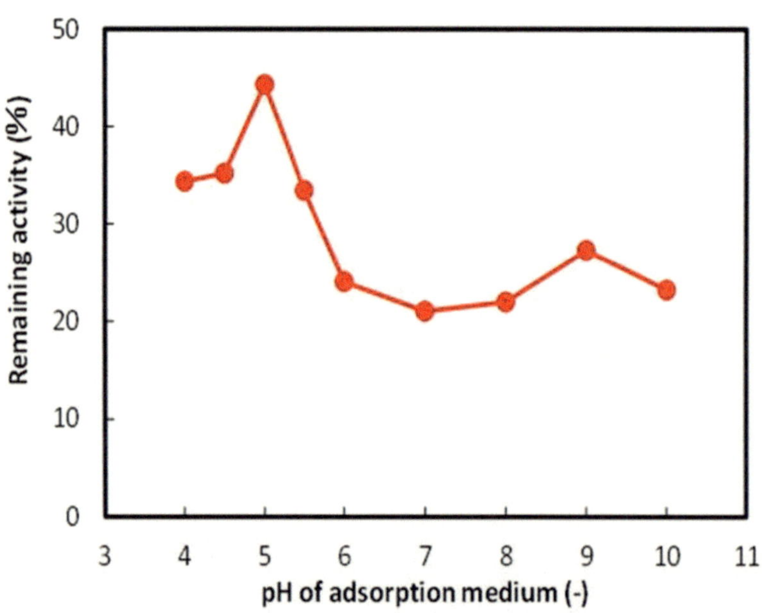

Figure 13.
Effect of pH of adsorption medium on the remaining activity of lysozyme adsorbed on bamboo charcoal after the heat treatment at 90°C for 30 min.

Three-dimensional structures of enzymes consist of secondary structures such as α-helix and β-sheet [28]. To elucidate the influence of adsorption on the structure of lysozyme, bamboo charcoal-adsorbed lysozyme has been measured by Fourier-transform infrared (FTIR) spectroscopy. **Figure 14** shows the FTIR spectra of native lysozyme and lysozyme adsorbed on bamboo at different pH. The most sensitive spectral region to enzyme secondary structural components is amide I (1700–1600 cm^{-1}), which is due almost entirely to the C=O stretch vibrations of peptide linkages [28]. The spectral pattern of bamboo charcoal-adsorbed lysozyme was influenced by the pH of adsorption medium. To evaluate the change in the secondary structure of bamboo-adsorbed lysozyme, the ratio of the absorbance at 1681 cm^{-1} to the absorbance at 1647 cm^{-1} (ABS_{1681}/ABS_{1647}) has been assessed since the band located at ca. 1681 cm^{-1} is assigned to intramolecular β-sheet and the band located at ca. 1647 cm^{-1} is assigned to α-helix. The ABS_{1681}/ABS_{1647} ratio at pH 5 (0.86), where the remaining activity showed the maximum value, was similar to that of native lysozyme (0.88). Likewise, the ABS_{1681}/ABS_{1647} ratio at pH 4 (0.92) was near that of native lysozyme. On the other hand, the ABS_{1681}/ABS_{1647} ratios at pH 7 (0.69) and 9 (0.61) were different from that of native lysozyme. The effect of solution pH of adsorption medium on the thermal stability of bamboo charcoal-adsorbed lysozyme has been summarized as follows. The structure of bamboo charcoal-adsorbed lysozyme was nearly the native structure of lysozyme when lysozyme was adsorbed on bamboo charcoal at pH 4, but the electrostatic interaction between lysozyme and bamboo charcoal could not sufficiently contribute to the thermal stability. The electrostatic interaction between lysozyme and bamboo charcoal could strongly retain the structure of lysozyme at high temperatures when lysozyme was adsorbed on bamboo charcoal at pH 5, where the native structure of lysozyme was maintained. The structure of lysozyme was partially destroyed since the electrostatic interaction was too strong to maintain the native structure of lysozyme, and the thermal stability of bamboo charcoal-adsorbed lysozyme dropped when lysozyme was adsorbed on bamboo charcoal at pH 7 and 9. Therefore, these results indicate that biochar-adsorbed enzymes exhibit the excellent thermal stability when the

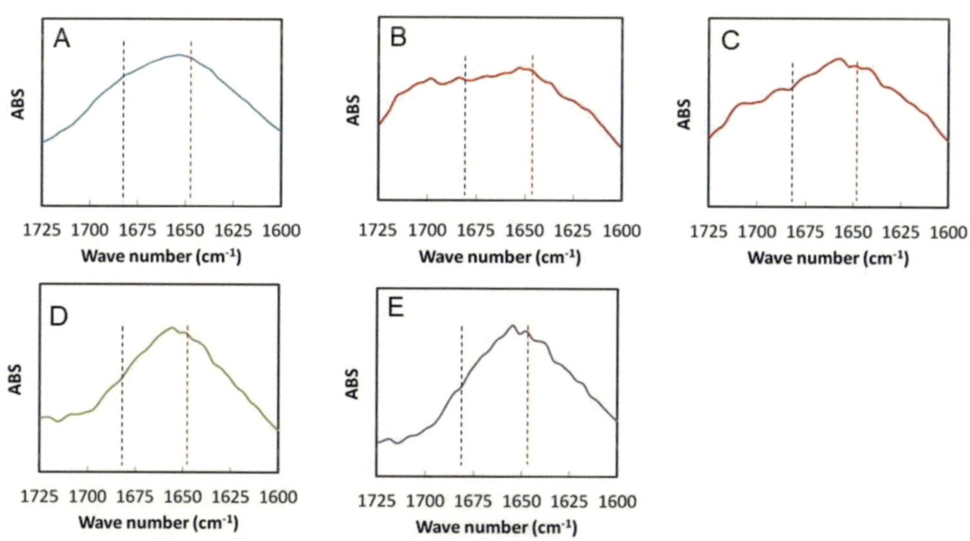

Figure 14.
(A) FTIR spectrum of native lysozyme. (B) FTIR spectrum of lysozyme adsorbed on bamboo charcoal at pH 4. (C) FTIR spectrum of lysozyme adsorbed on bamboo charcoal at pH 5. (D) FTIR spectrum of lysozyme adsorbed on bamboo charcoal at pH 7. (E) FTIR spectrum of lysozyme adsorbed on bamboo charcoal at pH 9.

native structure of enzymes is kept after the adsorption and the adsorption force is strong enough to retain the structure of enzymes against the heat stress.

2.4 Organic solvent stress tolerance of enzymes adsorbed on biochar

Biotransformation catalyzed by an enzyme in nonaqueous media has been applied to numerous synthetic processes because of the following benefits [29]: (1) The solubility of nonpolar reactants and products is improved. (2) Synthetic reactions can take place by the use of a conventional hydrolase without an expensive energy substance such as adenosine triphosphate (ATP). (3) The stereoselectivity of enzymes is markedly altered. (4) The thermal stability of enzymes is highly improved. (5) Enzymes can easily be recycled by the filtration or the centrifugation. (6) The product can easily be recovered by the evaporation when the volatile organic solvent is used as a reaction medium. (7) The contamination such as the growth of microorganisms can be inhibited by using organic solvents. However, the enzyme tends to show the low activity in organic solvents, compared to that in water since an organic solvent in general works as a denaturant of enzymes [30].

Figure 15 shows the scheme of reaction catalyzed by α-chymotrypsin (α-CT) [30, 31]. When *N*-acetyl-L-tyrosine ethyl ester (*N*-Ac-Tyr-OEt) is used as a substrate, the enzymatic reaction proceeds by the formation of enzyme intermediates between the active site of enzymes and the substrate. In water α-CT catalyzes the hydrolysis reaction of *N*-acetyl-L-tyrosine ethyl ester (*N*-Ac-Tyr-OEt) with water to give *N*-acetyl-L-tyrosine (*N*-Ac-Tyr-OH). On the other hand, in organic solvents α-CT mainly catalyzes the transesterification reaction of *N*-acetyl-L-tyrosine ethyl ester (*N*-Ac-Tyr-OEt) with another substrate, *n*-butanol (BuOH), to produce *N*-acetyl-L-tyrosine butyl ester (*N*-Ac-Tyr-OBu). Thus α-CT, which is hydrolase, can beneficially catalyze the synthetic reaction in organic solvents although expensive energy substances such as adenosine triphosphate (ATP) are needed for conventional synthetic reactions by enzymes to take place in water [24].

In order to evaluate the catalytic activity of enzymes adsorbed on biochar in an organic solvent, the transesterification catalyzed by α-CT adsorbed on the different kind of biochar in acetonitrile has been examined [19]. **Figure 16** shows the initial rates of *N*-acetyl-L-tyrosine butyl ester (*N*-Ac-Tyr-OBu) and *N*-acetyl-L-tyrosine (*N*-Ac-Tyr-OH) catalyzed by free and biochar-adsorbed α-CT in acetonitrile containing 5% (v/v) water. Both initial rates of *N*-Ac-Tyr-OBu and *N*-Ac-Tyr-OH catalyzed by biochar-adsorbed α-CT were much higher than those catalyzed by free α-CT. α-CT adsorbed on bamboo charcoal was the most effective of all, with respect

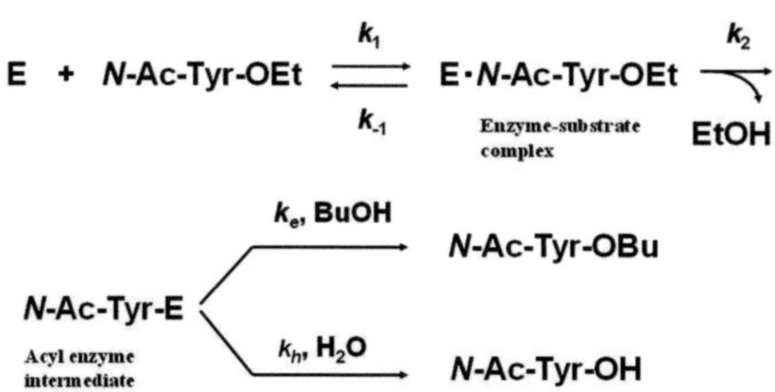

Figure 15.
α-CT-catalyzed transesterification of N-acetyl-L-tyrosine ethyl ester (N-Ac-Tyr-OEt) with n-butanol (BuOH) to N-acetyl-L-tyrosine butyl ester (N-Ac-Tyr-OBu) and competing hydrolysis (N-Ac-Tyr-OH).

to the enhancement in the initial rates of *N*-Ac-Tyr-OBu and *N*-Ac-Tyr-OH. With regard to the transesterification catalyzed by hydrolase, which is characteristic of the nonaqueous enzymology, the initial rate of *N*-Ac-Tyr-OBu catalyzed by α-CT adsorbed on bamboo charcoal was about 50 times higher than that catalyzed by free one. Enzymes are aggregated in an organic solvent, and most of them cannot directly come in contact with the bulk organic phase containing substrates, although they are soluble in an aqueous solution. On the other hand, most of the enzymes adsorbed on biochar are directly in contact with the bulk organic phase since they are located on the surface of biochar. Accordingly, biochar-adsorbed enzymes can effectively proceed with the reaction, compared to free enzymes, since mass transfer of substrates and products is rapidly facilitated [32].

In order to elucidate the influence of a kind of biochar on the secondary structure of α-CT, the FTIR spectra of free and biochar-adsorbed α-CT were measured. **Table 4** shows the ratio of the absorbance at 1650 cm^{-1} to the absorbance at 1630 cm^{-1} (ABS_{1650}/ABS_{1630}) of free α-CT and α-CT adsorbed onto biochar. As mentioned above, the band located at ca. 1650 cm^{-1} is assignable to α-helix, and the band located at ca. 1630 cm^{-1} is assignable to intramolecular β-sheet. The order of the ABS_{1650}/ABS_{1630} ratio was bamboo charcoal-adsorbed α-CT > adzuki bean charcoal-adsorbed α-CT = wood charcoal-adsorbed α-CT > free α-CT. The order of the ABS_{1650}/ABS_{1630} ratio was similar to that of the initial rate of transesterification as shown in **Figure 16**. The α-helical structure of α-CT molecule is more changeable than β-sheet, since the β-sheet structure is the main backbone of α-CT molecule [33]. Thus, the results indicate that at the higher initial rate, the transesterification is catalyzed by α-CT molecules having the secondary structure kept more highly.

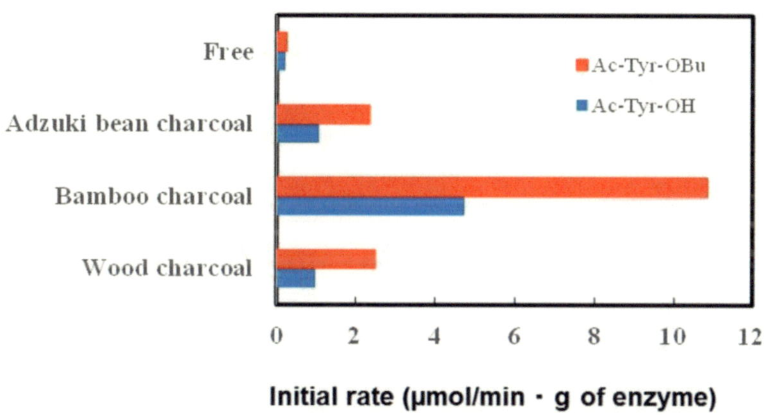

Figure 16.
Dependence of kind of biochar on biochar-adsorbed α-CT-catalyzed transesterification. Free or biochar-adsorbed α-CT was placed in acetonitrile containing 5% (v/v) water, 10 mM N-Ac-Tyr-OEt, 1000 mM BuOH, and 1 mM acetanilide, and the resulting mixture was shaken at 120 rpm and 25°C.

Sample	ABS_{1650}/ABS_{1630} (—)
Free α-CT	1.1
Adzuki bean charcoal-adsorbed α-CT	1.3
Bamboo charcoal-adsorbed α-CT	1.5
Wood charcoal-adsorbed α-CT	1.3

Table 4.
Ratio of the absorbance at 1650 cm^{-1} to the absorbance at 1630 cm^{-1} (ABS_{1650}/ABS_{1630}) of α-CT provided by the FTIR measurement.

Moreover, it is suggested that the content of functional groups in bamboo charcoal is suitable to keep the secondary structure of α-CT in acetonitrile since functional groups contribute to the adsorption of α-CT on biochar.

Figure 17 shows time course of remaining activities of free α-CT and bamboo charcoal-adsorbed α-CT through the heat treatment at 50°C [20]. The state of free α-CT in acetonitrile, where α-CT was dispersed as the solid state, was unchanged during the heat treatment, although enzymes dissolved in an aqueous solution immediately form the aggregation of thermally denatured enzymes [15]. Likewise, the enzyme aggregation and the cohesion among bamboo charcoal-adsorbed α-CT were not observed in acetonitrile during the heat treatment. However, the remaining activities of free α-CT and BCP-adsorbed α-CT gradually dropped with an increase in heat time. The relation between the remaining activity of free α-CT and heat time could be correlated by first-order kinetics, while the relation between bamboo charcoal-adsorbed α-CT and heat time could be correlated by second-order kinetics. When the curve fitting was carried out in the figure, the half-life of inactivation of free α-CT was 33 min, and the half-life of inactivation of bamboo charcoal-adsorbed α-CT was 125 min. Therefore, the half-life of bamboo charcoal-adsorbed α-CT showed 3.8-fold, compared with that of α-CT. On the other hand, the half-life of inactivation of bamboo charcoal-adsorbed α-CT is 15 min in aqueous solutions at 45°C [17]. Consequently, the thermal stability of bamboo charcoal-adsorbed α-CT in acetonitrile was greater than that in water. As a result, the electrostatic interaction between α-CT and bamboo charcoal, which mainly contributes to the adsorption of α-CT on bamboo charcoal, is strengthened in acetonitrile as the dielectric constant of acetonitrile is much smaller than that of water [34].

In moist air, the catalytic activity of solid enzymes is markedly dependent on the thermodynamic water activity (a_w), which is defined as the ratio of the water partial pressure to the vapor pressure of pure water [35]. Similarly, the catalytic activity of biochar-adsorbed enzymes might be influenced in hydrophilic organic solvents containing low water content. **Figure 18** shows the relationship of the initial transesterification rate (V_e) and the initial hydrolysis rate (V_h) with the water activity

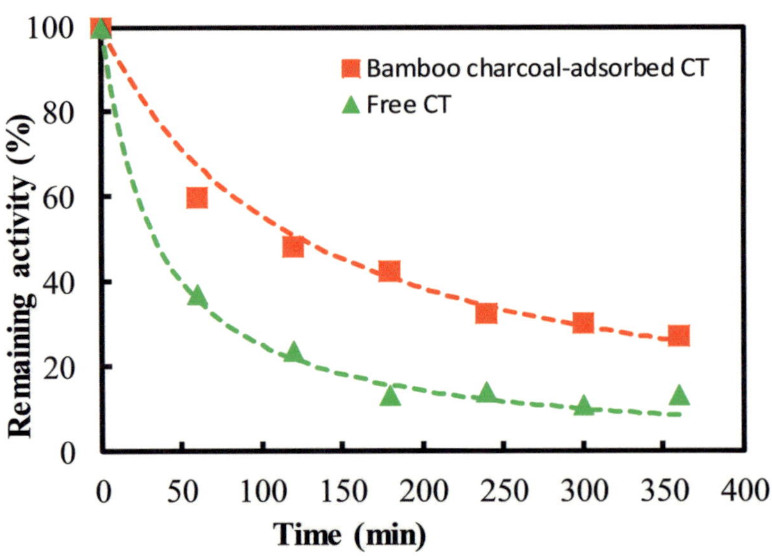

Figure 17.
Time course of remaining activities of free α-CT and bamboo charcoal-adsorbed α-CT through the heat treatment at 50°C.

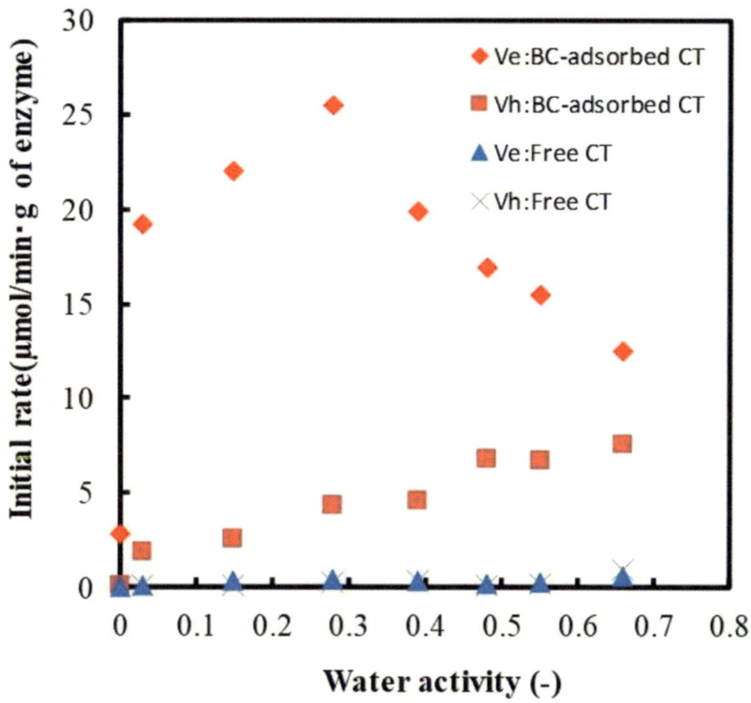

Figure 18.
Effect of water activity on the transesterification rate (V_e) and hydrolysis rate (V_h) of free α-CT and bamboo charcoal (BC)-adsorbed α-CT in acetonitrile.

in acetonitrile at 25°C [20]. Low water activity inhibited the inherent enzymatic hydrolysis of *N*-acetyl-L-tyrosine ethyl ester (*N*-Ac-Tyr-OEt) with water, whereas the enzymatic transesterification of *N*-acetyl-L-tyrosine ethyl ester (*N*-Ac-Tyr-OEt) with *n*-butanol (BuOH) was enhanced. The water activity strongly affected the initial transesterification rates catalyzed by bamboo charcoal-adsorbed α-CT and free α-CT, and the correlation between these parameters showed a bell-shaped curve. The initial transesterification catalyzed by bamboo charcoal-adsorbed α-CT exhibited about 60-fold, compared to that catalyzed by free α-CT when the maximum initial transesterification rates of free α-CT and bamboo charcoal-adsorbed α-CT were shown. On the other hand, the initial hydrolysis rates catalyzed by free α-CT and bamboo charcoal-adsorbed α-CT gradually increased with increasing the water activity. The relationship of the catalytic activity of enzymes with the water activity in organic solvents tends to depict a bell-shaped curve. The balance between the kinetic rigidity of enzyme structures and their thermodynamic stability results in the optimum water activity [29, 36]. The kinetic rigidity drops with an increase in water activity, while the native structure of enzymes is gradually influenced by the thermodynamic stability. Consequently, the catalytic activity of enzymes increases with increasing the flexibility in the rigid structure of enzymes and then decreases with increasing the disturbance of enzyme structures. On the other hand, the hydrolysis is promoted with an increase in the water activity since the high water activity results in the high overall water concentration. Concerning the reaction selectivity, which was defined as the ratio of the initial transesterification rate (V_e) to the initial hydrolysis rate (V_h), bamboo charcoal-adsorbed α-CT was much better than free α-CT.

Table 5 shows the absorbance ratio at 1650 and 1630 cm^{-1} (ABS_{1650}/ABS_{1630}) of bamboo charcoal-adsorbed α-CT [20]. The higher the absorbance ratio, the higher the secondary structure. The water activity did not affect the absorbance ratio

Water activity (−)	ABS_{1650}/ABS_{1630} (−)
0.03	1.3
0.28	1.3
0.55	1.3
0.73	1.3

Table 5.
Ratio of the absorbance at 1650 cm^{-1} to the absorbance at 1630 cm^{-1} (ABS_{1650}/ABS_{1630}) of bamboo charcoal-adsorbed α-CT provided by the FTIR measurement.

(ABS_{1650}/ABS_{1630}) of bamboo charcoal-adsorbed α-CT, indicating that the secondary structure of bamboo charcoal-adsorbed α-CT does not depend on the water activity. Accordingly, the adsorption firmly makes the conformation of bamboo charcoal-adsorbed α-CT maintained. The absorbance ratio (ABS_{1650}/ABS_{1630}) of bamboo charcoal-adsorbed α-CT is higher than that of free α-CT, as seen in **Table 4**. The results illustrate that the water activity effectively affects the catalytic activity of bamboo charcoal-adsorbed α-CT having a native structure, compared to that of free α-CT.

The catalysis of free α-CT and bamboo charcoal-adsorbed α-CT was markedly dependent upon the nature of organic solvents as shown in **Figure 19**. The catalytic activity of bamboo charcoal-adsorbed α-CT was much superior to that of free α-CT in organic solvents. The initial transesterification rate of free α-CT in *n*-octane exhibited 813-fold, compared to that in *n*-butyl acetate, while the initial transesterification rate of bamboo charcoal-adsorbed α-CT in *n*-octane was 1.3 times greater than that in *n*-butyl acetate.

There have been some reports that the native conformation of enzymes may be altered when enzymes are immersed in organic solvents [37, 38]. **Table 6** shows the ratio of the absorbance at 1650 cm^{-1} to the absorbance at 1630 cm^{-1} (ABS_{1650}/ABS_{1630}) of free α-CT and bamboo charcoal-adsorbed α-CT after they were immersed in organic solvents for 24 h. The absorbance ratio (ABS_{1650}/ABS_{1630}) of bamboo charcoal-adsorbed α-CT after the solvent immersion was similar to that before the solvent immersion. On the other hand, the absorbance ratio (ABS_{1650}/ABS_{1630}) of free α-CT was altered by the immersion in octane. Those results indicate that the conformation of bamboo charcoal-adsorbed α-CT is hardly influenced by the nature of solvents, compared to the case of free α-CT.

Figure 20 shows the relation between the catalytic activity and the hydrophobicity defined as log P where P is a partition coefficient for a given solvent between *n*-octanol and water [39]. The initial transesterification rate of free α-CT and bamboo charcoal-adsorbed α-CT increased with an increase in the hydrophobicity of organic solvents. Likewise, the initial hydrolysis rates of free α-CT and bamboo charcoal-adsorbed α-CT tended to increase with increasing the hydrophobicity of organic solvents. The hydrophobicity contributes to the partition of water between enzyme molecules and the bulk organic phase in reaction systems [29, 40]. When a certain amount of water is added into organic solvents, the amount of water associated with enzyme molecules increases in an increase in the hydrophobicity of organic solvents. Thereby, the flexibility of enzyme molecules, which is essential for catalytic activity, is enhanced by the hydration of enzyme molecules, and then the catalytic activity increases. On the other hand, hydrolysis reaction is promoted by the increase of water molecules around enzymes. As shown in **Figure 20**, the initial transesterification rate of bamboo charcoal-adsorbed α-CT was much superior to that of free α-CT at the low hydrophobicity. For instance, the initial transesterification rate of bamboo charcoal-adsorbed α-CT was about 760 times higher than that

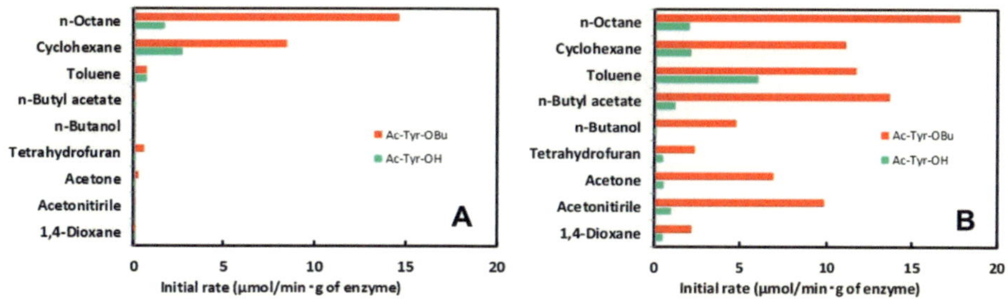

Figure 19.
Solvent dependence o of transesterification catalyzed by free α-CT (A) and BCP-adsorbed α-CT (B).

Solvent	ABS_{1650}/ABS_{1630} (−)	
	Free α-CT	**Bamboo charcoal-adsorbed α-CT**
None	1.14	1.47
Acetonitrile	1.15	1.51
n-Octane	1.20	1.53

Table 6.
Ratio of the absorbance at 1650 cm^{-1} to the absorbance at 1630 cm^{-1} (ABS_{1650}/ABS_{1630}) of free α-CT and bamboo charcoal-adsorbed α-CT provided by the FTIR measurement after the solvent immersion.

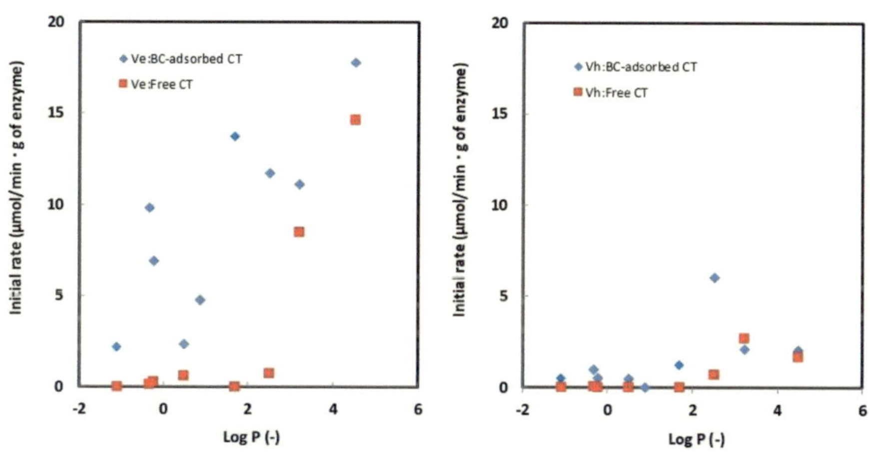

Figure 20.
Relationship of log P of solvents with transesterification rate (A) or hydrolysis rate (B) of free α-CT and bamboo charcoal (BC)-adsorbed α-CT.

of free α-CT in *n*-butyl acetate (log P = 1.7). Moreover, bamboo charcoal-adsorbed α-CT depicted the high initial transesterification rate in acetone and acetonitrile, compared to the case of free α-CT. An enzymatic reaction in hydrophilic solvents has the advantage of the solubility of a variety of substrates including drug derivatives, which are poorly soluble in hydrophobic solvents [41]. As discussed above, since the native conformation of bamboo charcoal-adsorbed α-CT was maintained in hydrophilic organic solvents, bamboo charcoal-adsorbed α-CT could exhibit the high catalytic activity, compared to the case of free α-CT. On the other hand, no correlation between the catalytic activity and the other parameter (e.g., dielectric constant, hydrogen bonding parameter, solubility parameter, and viscosity) was shown.

3. Conclusions

This chapter has introduced the study on the application of the biochar to enzyme carriers to develop the high value-added application of biomass materials. Biochar was thermal stable, chemical stable, insoluble under reaction conditions, available, low cost, regeneration, and reusable. Moreover, as biochar had functional groups for the interaction of enzymes and a high affinity to enzymes, enzymes were firmly adsorbed on biochar. On the other hand, the original weakness of enzymes due to the heat and organic solvent stresses could be much improved by adsorbing enzymes onto the biochar. Moreover, enzymes are strictly influenced not only by heat and organic solvents but also by ultraviolet, X-ray, sound wave, shake, freeze, pressure, shearing force, extreme ionic strength, urea, surfactant metal ion, reductant, and so on. It would be expected that suitable carriers having the high tolerance against those stresses are developed for an enzyme carrier by selecting a kind of biochar as well because there are a great variety of biomass in the earth.

Acknowledgements

This work was mainly supported by a Grant-in-Aid for Scientific Research (C) from Japan Society for the Promotion of Science (No. 24561013) and a Grant-in-Aid for Scientific Research from Japan Science and Technology Agency (No. AS2111014D).

Author details

Hidetaka Noritomi
Department of Applied Chemistry for Environment, Tokyo Metropolitan University, Tokyo, Japan

*Address all correspondence to: noritomi@tmu.ac.jp

References

[1] Olivier JGJ, Muntean M, Peters JAHW. Trends in Global CO_2 Emissions. 2017 Report; 2017

[2] IPCC. Climate Change 2014: Synthesis Report. Switzerland: Intergovernmental Panel on Climate Change (IPCC); 2014

[3] Ho YC, Show KY. A perspective in renewable energy production from biomass pyrolysis—Challenges and prospects. Current Organic Chemistry. 2015;**19**:423-436

[4] Straathof AJJ. Transformation of biomass into commodity chemicals using enzymes or cells. Chemical Reviews. 2014;**114**:1871-1908

[5] Buchholz K, Kasche V, Bornscheuer UT. Biocatalyst and Enzyme Technology. 2nd ed. Weinheim: Wiley-Blackwell; 2012

[6] Silwana B, Horst CVD, Iwuoha E. Aerometric determination of cadmium, lead, and mercury metal ions using a novel polymer immobilized horseradish peroxidase biosensor system. Journal of Environmental Science and Health, Part A. 2014;**49**:1501-1511

[7] Leech D, Kavanagh P, Schuhmann W. Enzymatic fuel cells: Recent progress. Electrochimica Acta. 2012;**84**:223-234

[8] Bailey JE, Ollis DF. Biochemical Engineering Fundamentals. 2nd ed. New York: McGraw-Hill; 1986

[9] Zdata J, Meyer AS, Jesionowski T, Pinelo M. A general overview of support materials for enzyme immobilization: Characteristics, properties, practical utility. Catalysts. 2018;**8**:92-118

[10] Mateo C, Palomo JM, Fernandez-Lorente G, Guisan JM, Fernandez-Lorente R. Improvement of enzyme activity, stability and selectivity via immobilization techniques. Enzyme and Microbial Technology. 2007;**40**:1451-1463

[11] Elnashar MMM. Review article: Immobilized molecules using biomaterials and nanobiotechnology. Journal of Biomaterials and Nanobiotechnology. 2010;**1**:61-77

[12] Anderson CR, Condron LM, Clough TJ, Fiers M, Alison S, Hill RA, et al. Biochar induced soil microbial community change: Implications for biogeochemical cycling of carbon, nitrogen and phosphorus. Pedobiologia. 2011;**54**:309-320

[13] Noritomi H, Iwai D, Kai R, Tanaka M, Kato S. Adsorption of lysozyme on biomass charcoal powder prepared from plant biomass wastes. Journal of Chemical Engineering of Japan. 2013;**46**:196-200

[14] Noritomi H, Hishinuma K, Kurihara S, Nishigami J, Takemoto T, Endo N, et al. Adsorption of α-chymotrypsin on plant biomass charcoal. Journal of Surface Engineered Materials and Advanced Technology. 2013;**3**:269-274

[15] Noritomi H, Kai R, Iwai D, Tanaka H, Kamila R, Tanaka M, et al. Increase in thermal stability of proteins adsorbed on biomass charcoal powder prepared from plant biomass wastes. Journal of Biomedical Science and Engineering. 2011;**4**:692-698

[16] Noritomi H, Ishiyama R, Kai R, Iwai D, Tanaka M, Kato S. Immobilization of lysozyme on biomass charcoal powder derived from plant biomass wastes. Journal of Biomaterials and Nanobiotechnology. 2012;**3**:446-451

[17] Noritomi H, Kurihara S, Endo N, Kato S. Heat-resistant properties of α-chymotrypsin adsorbed onto biomass

charcoal powder. Journal of Biomaterials and Nanobiotechnology. 2014;**5**:179-185

[18] Noritomi H, Kurihara S, Endo N, Kato S, Uchiyama K. Effect of adsorption condition on thermal stability of proteins adsorbed onto biomass charcoal powder. International Journal of GEOMATE. 2016;**11**:2123-2128

[19] Noritomi H, Nishigami J, Endo N, Kato S, Uchiyama K. Organic solvent-resistant properties of proteins adsorbed onto biomass charcoal powder. International Journal of GEOMATE. 2017;**12**:140-145

[20] Noritomi H, Nishigami J, Endo N, Kato S, Uchiyama K. Influence of water activity on protease adsorbed on biochar in organic solvents. Journal of Materials Science Research. 2017;**6**:96-102

[21] Noritomi H, Nishigami J, Endo N, Kato S, Takagi S. Effect of solvent on catalysis of protease adsorbed on biochar in organic media. Journal of Materials Science Research. 2018;**7**:46-52

[22] Asada T, Ishihara S, Yamane T, Toba A, Yamada A, Oikawa K. Science of bamboo charcoal: Study on carbonizing temperature of bamboo charcoal and removal capability of harmful gases. Journal of Health Science. 2002;**48**:473-479

[23] Nishimiya K, Hata T, Imamura Y, Ishihara S. Analysis of chemical structure of wood charcoal by X-ray photoelectron spectroscopy. Journal of Wood Science. 1998;**44**:56-61

[24] Adamson AW. Physical Chemistry of Surfaces. 4th ed. New York: John Wiley & Sons; 1982

[25] Creighton TE. Protein Function: Practical Approach. Oxford: IRL Press; 1989

[26] Nohara D, Mizutani A, Sakai T. Kinetic study on thermal denaturation of hen egg-white lysozyme involving precipitation. Journal of Bioscience and Bioengineering. 1999;**87**:199-205

[27] Noritomi H, Minamisawa K, Kamiya R, Kato S. Thermal stability of proteins in the presence of aprotic ionic liquids. Journal of Biomedical Science and Engineering. 2011;**4**:94-99

[28] Surewicz WK, Mantsch HH. New insight into protein secondary structure from resolution-enhanced infrared spectra. Biochimica et Biophysica Acta. 1988;**952**:115-130

[29] Klibanov AM. Improving enzymes by using them in organic solvents. Nature. 2001;**409**:241-246

[30] Lehninger AL, Nelson DL, Cox MM. Principles of Biochemistry. 2nd ed. New York: Worth; 1993

[31] Wescott CR, Noritomi H, Klibanov AM. Rational control of enzymatic enantioselectivity through solvation thermodynamics. Journal of the American Chemical Society. 1996;**118**:10365-10370

[32] Fogler HS. Elements of Chemical Reaction Engineering. 2nd ed. New Jersey: Prentice Hall PTR; 1992

[33] Kumar A, Venkatesu P. Overview of the stability of α-chymotrypsin in different solvent media. Chemical Reviews. 2012;**112**:4283-4307

[34] Reinhardt C. Solvents and Solvent Effects in Organic Chemistry. 2nd ed. Weinheim: Wiley-VCH; 1988

[35] Acker L. Enzyme reactions in foods of low moisture content. Advances in Food Research. 1962;**11**:263-330

[36] Bell G, Janssen AEM, Halling PJ. Water activity fails to predict critical hydration level for enzyme activity in

polar organic solvents: Interconversion of water concentrations and activities. Enzyme and Microbial Technology. 1997;**20**:471-477

[37] Mozhaev VV, Khmelnitsky YL, Sergeeva MV, Belova AB, Klyachko NL, Levashov AV, et al. Catalytic activity and denaturation of enzymes in water/organic cosolvent mixtures. α-Chymotrypsin and laccase in mixed water/alcohol, water/glycol and water/formamide solvents. European Journal of Biochemistry. 1989;**184**:597-602

[38] Ryu K, Dordick JS. How do organic solvents affect peroxidase structure and function? Biochemistry. 1992;**31**:2588-2598

[39] Laane C, Boeren S, Vos K, Veeger C. Rules for optimization of biocatalysis in organic solvents. Biotechnology and Bioengineering. 1987;**30**:81-87

[40] Zaks A, Klibanov AM. Enzymatic catalysis in nonaqueous solvents. The Journal of Biological Chemistry. 1988;**263**:3194-3201

[41] Kise H, Hayakawa H, Noritomi H. Protease-catalyzed synthetic reactions and immobilization-activation of the enzymes in hydrophilic organic solvents. Journal of Biotechnology. 1990;**14**:239-254

Chapter 6

Biochar Phosphorus Sorption-Desorption: Potential Phosphorus Eutrophication Mitigation Strategy

Lucy W. Ngatia, Johnny M. Grace III, Daniel Moriasi, Alejandro Bolques, George K. Osei and Robert W. Taylor

Abstract

Phosphorus (P) eutrophication in the water bodies is of global concern. The role of biochar in the mitigation of (P) eutrophication has recently received substantial attention. Agriculture is the main source of P in the water bodies, as a result of excessive fertilizer and manure application. Excessive P results in excessive primary production in the water bodies, leading to anoxic conditions, growth of toxic algae blooms, altering plant species composition and biomass. Therefore, resulting in food web disruption, fish kill, toxins production and recreation areas degradation. When biochar is applied on farm, it has potential to sorb/adsorb P, immobilizing it, slowing its translocation to the water bodies. However, biochar effectiveness in P sorption is influenced by both feedstock type and pyrolysis temperature. The interaction between feedstock type and pyrolysis temperature influences the biochar pH, surface area, aromatic carbon, cation exchange capacity, surface charge density, biochar internal porosity and polar and nonpolar surface sites that promote nutrient absorption. Hence, biochar properties have a broad spectrum that influences how biochar reacts with P sorption; therefore, it is not appropriate to extrapolate observed results to different materials. Biochar that promote P sorption rather than desorption should be considered and designed to meet specific management practices.

Keywords: biochar, desorption, eutrophication, phosphorus, mitigation, sorption

1. Introduction

Phosphorus (P) eutrophication is a major problem globally. Increased aquatic primary production attributed to P over enrichment results in eutrophication [1, 2]. Aquatic systems affected by eutrophication often exhibit harmful algal blooms, which foul water intakes and waterways, disrupt food webs, fuel hypoxia and produce secondary metabolites that are toxic to water consumers and users including human, cattle, zooplankton, shellfish, domestic pets and fish [3]. Agriculture is the main pollutant of aquatic systems [4]. Overapplication of P fertilizer and manure to soil is in itself wasteful and causes P accumulation in the soil, but the transport of P to aquatic ecosystems by erosion is also causing widespread problems of eutrophication [5–8]. Soil P exists in both organic and inorganic forms [9]; the inorganic P is highly influenced by soil pH and is mainly coupled with amorphous and crystalline forms of Fe, Al, and Ca [10]. Biochar exhibits potential to mitigate P eutrophication when applied on farm as a result of its high capacity to sorb P [1]. Biochar is a by-product of pyrolysis process, whereby in an energy-limited world, biomass is converted to energy products through pyrolysis process [11]. Biochar nutrient sorption capacity has been reported to exceed that of natural organic matter by a factor of 10–100 in some cases [12]. However, its capacity for P sorption is influenced by both feedstock type and pyrolysis temperature [13]. The interaction between feedstock type and pyrolysis temperature influences biochar characteristics resulting in wide spectrum of biochar properties that influence P sorption [1, 14]. Therefore, extrapolation of P sorption findings from one material to another is not appropriate; as a result of diversity of biochar response to P sorption, every material needs to be studied and designed to address the set P sorption objectives.

2. Forms of phosphorus

Soil P exists as either organic or inorganic compounds that differ significantly in their biological availability in the soil environment [9]. Phosphorus goes through different geochemical processes in soil which include adsorption, solubilization, precipitation and complexation that determine its mobility and fate [15]. The inorganic P compounds mainly couple with amorphous and crystalline forms of Fe, Al, and Ca [10] and are highly influenced by soil acidity and alkalinity [1]. Soil acidity and alkalinity influence the impact of solubilization, complexation, adsorption and precipitation on P retention and release. In acidic soils P is fixed by sorption to oxides and hydroxides of Al and Fe through formation of insoluble Al and Fe phosphates by ligand exchange and precipitation reactions [16, 17]. In alkaline soils, P is made unavailable due to formation of metal complexes such as Mg-P and Ca-P [18, 19]. The organic P in most soils is dominated by a mixture of phosphate monoesters (e.g., inositol phosphates and mononucleotides) and phosphate diesters (mainly phospholipids and nucleic acids), with smaller amounts of phosphonates (compounds with a direct carbon-phosphorus bond) and organic polyphosphates (e.g., adenosine triphosphate) [20]. Plants can acquire P from organic compounds through various mechanisms; some of the mechanisms allow plants to utilize organic P as efficiently as inorganic phosphate [21, 22]. Different environmental conditions influence availability and sorption of P, for example anaerobic condition leads to the release of P resulting from reduction of ferric to ferrous iron [23]. The presence of sulfate could lead to reaction of ferric iron with sulfate and sulfide to form ferrous iron and iron sulfide resulting in the release of P [24]. Increased temperature can reduce adsorption of P by mineral complexes in the sediment [25]. Other physicochemical processes affecting the release of P from the sediment include redox, temperature, reservoir hydrology, pH potential and environmental conditions [26]. These processes are further complicated by the influence of biological processes, for example mineralization, leading to a complex system regulating the release of P across sediment water interface [26]. Understanding of P retention and release mechanism enhances effective management of P enhancing crop production and promoting sustainability of soil and water quality [11].

3. Agriculture: the major source of phosphorus to water bodies

Increasing human population calls for increased food production. Increased food production requires increased fertilizer application which includes P fertilizer

(**Figure 1**) [27]. Today agriculture is heavily dependent on P derived from phosphate rock. Phosphorus is a nonrenewable resource and it is expected that economically mineable P reserves will be exhausted within 50–100 years (**Figure 2**) [27]. Crop and livestock production systems are the major cause of human alteration of the global P cycles [28]. The major source of P input to water bodies is the excessive application of fertilizer or manure on farm which causes P accumulation in soils [5]. Excess P or poor-timed application could mobilize P through runoff, negatively impacting water quality of water bodies and causing eutrophication [29, 30].

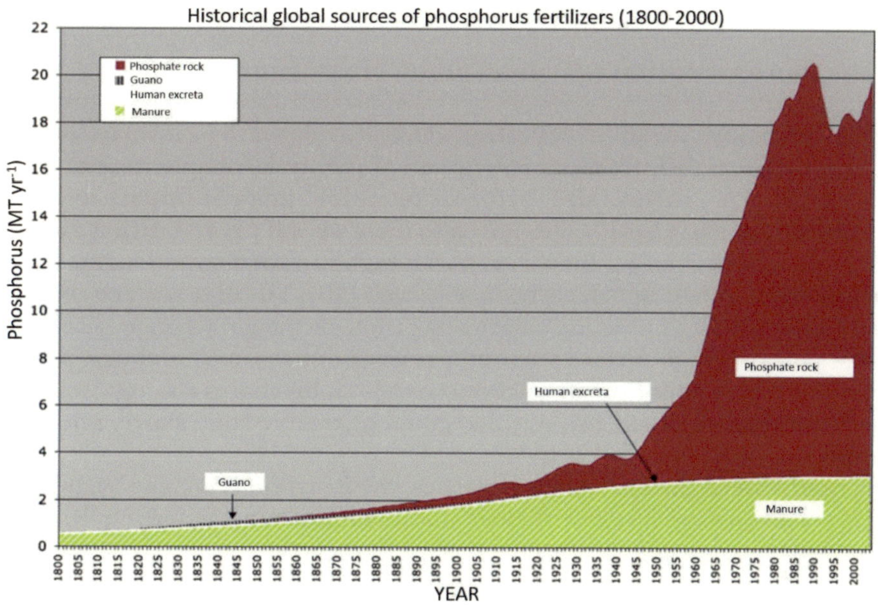

Figure 1.
Historical sources of phosphorus for use as fertilizers, including manure, human excreta, guano and phosphate rock (1800–2000). Modified from [27].

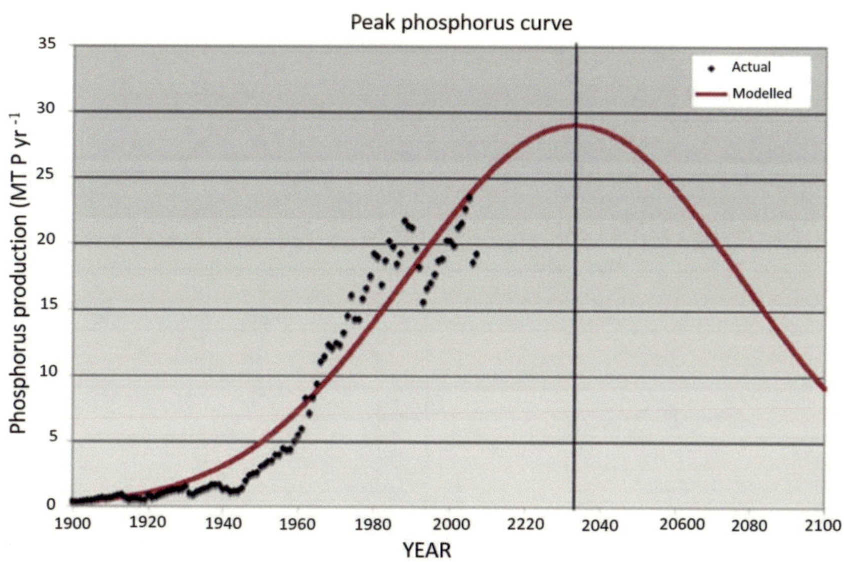

Figure 2.
Indicative peak phosphorus curve, illustrating that, global phosphorus reserves are likely to peak after which production will be significantly reduce. Modified from [27].

In many cases, P enters aquatic ecosystems sorbed to soil particles that are eroded into rivers, lakes and streams; any factor elevating soil P concentration or accelerating soil erosion potentially increase P runoff to aquatic system [31, 32]. Mobilization of P involves chemical, biochemical and biological processes. The processes are grouped into detachment or solubilization mechanisms and are defined by the physical size of the P compounds that are mobilized [33]. Potential for solubilization increases with increasing concentrations for extractable P. Depending on site conditions, diffuse P transport occurs as particulate or dissolved P in overland flow, drainage, channelized surface runoff or groundwater [34].

4. Phosphorus sorption is dependent on biochar properties

Biochar exhibits variable properties which are related to its formation; biochar properties are mainly influenced by method of pyrolysis, temperature of pyrolysis, feedstock type, particle size, pyrolysis retention time and furnace oxygen level [35–38]. Biochar has been reported to increase soil pH and cation exchange capacity (**Figure 3**) [1, 39]. Biochars exhibit high surface charge density, and CEC values up to 112 cmolc kg^{-1} have been observed [40]. The high surface charge density allows biochar to retain cations by cation exchange, whereas biochar internal porosity, high surface area and presence of both polar and non-polar surfaces sites promote absorption of nutrient [41]. Biochar is also reported to contain essential nutrients [41, 42] that could be retained and slowly released over time [43, 44].

As a result of biochar variable properties, the P sorption findings by different studies are inconsistent; some previous studies indicated no increase in P sorption

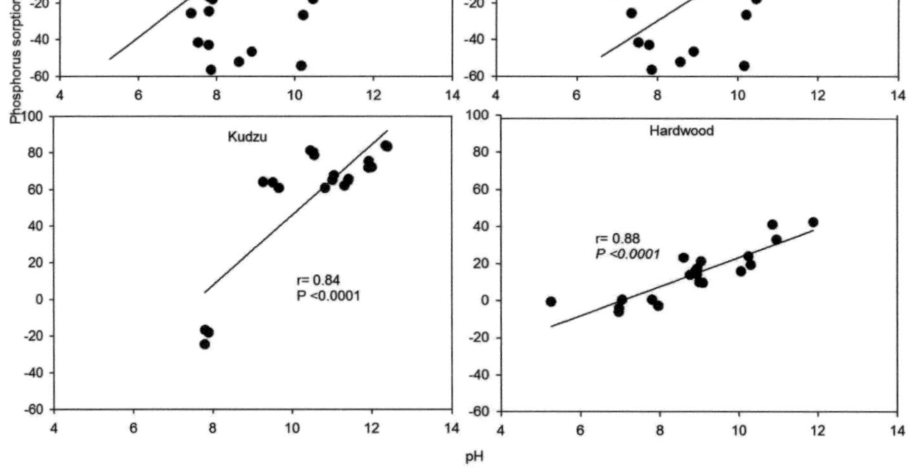

Figure 3.
Relationship between biochar phosphorus sorption and pH for all feedstocks, switchgrass, kudzu and Chinese tallow. Modified from [1].

of the sandy and clayey soils after addition of biochar [45]. There was no P sorption observed with corn biochar produced at 350 and 550°C pyrolysis temperatures [46]. Biochar produced from sugarcane bagasse, peanut hull, Brazilian pepper wood and bamboo at 200, 450 and 600°C did not indicate any clear trend in phosphorus sorption [47]. However, other studies indicated reduction of P leaching after biochar application [41, 48]. Further, biochar addition reduced P leaching after manure addition and in green roof discharge water [41, 49]. Application of 2% biochar to agricultural soil amended with swine manure resulted in reduction of P leaching by 69% [41]. Addition of 2% pecan shell biochar to loamy soil reduced P leachate by 40% [50]. Orange peel pyrolyzed at between 250 and 700°C improved P sorption by 8–83% [51]. [1] indicated that biochar pyrolyzed from kudzu and Chinese tallow at 300–700°C temperature exhibited increased P sorption; when switchgrass was pyrolyzed at 300–500°C, it exhibited P desorption, but when pyrolyzed at 500–700°C temperature, it exhibited P sorption, clearly demonstrating that feedstock type and pyrolysis temperature are major determinant of P sorption capacity (**Figures 4** and 5).

In [1], it is also demonstrated that increasing temperature resulted in loss of O-alkyl carbon and accumulation of aromatic carbon that favored P sorption (**Figure 6**). This variability in biochar capacity to sorb P suggests that understanding each biochar type is important to ensure appropriate application to meet target objectives.

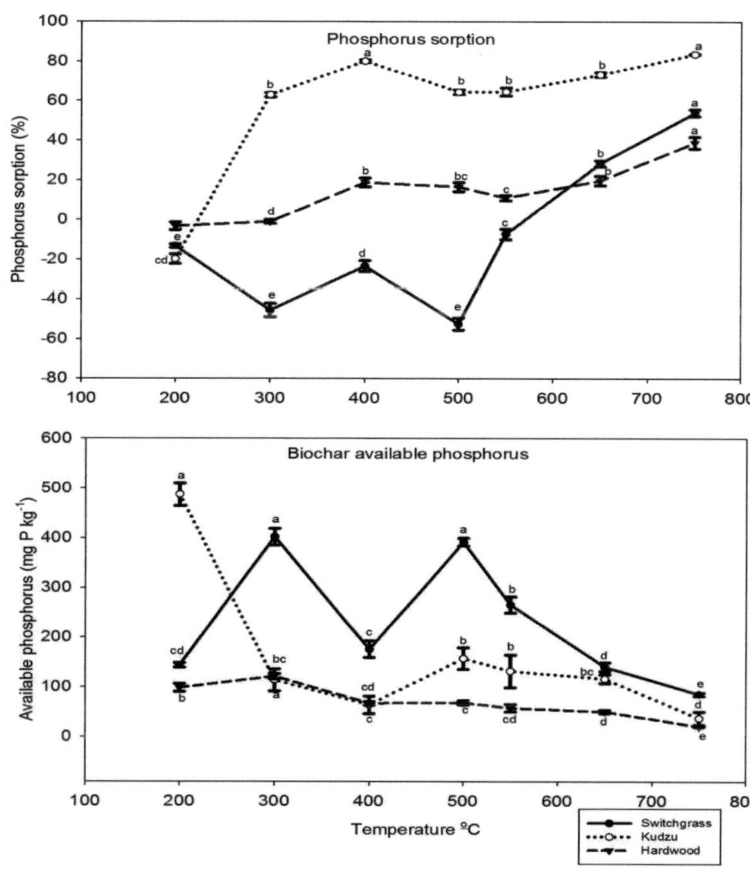

Figure 4.
Biochar phosphorus adsorption; Modified from [1].

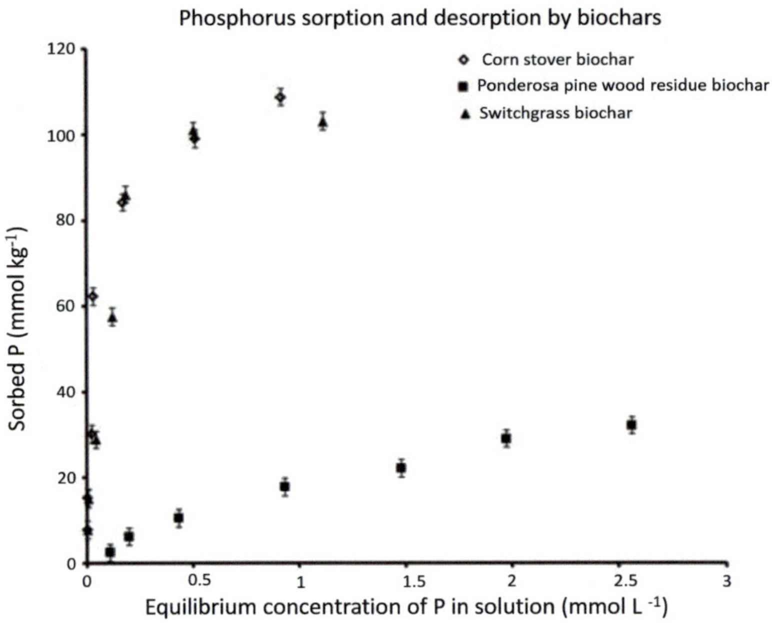

Figure 5.
Phosphorus sorption by corn stover, Ponderosa pine wood residue, and switchgrass biochars. Each data point is the mean of four replications with standard error. Modified from [11].

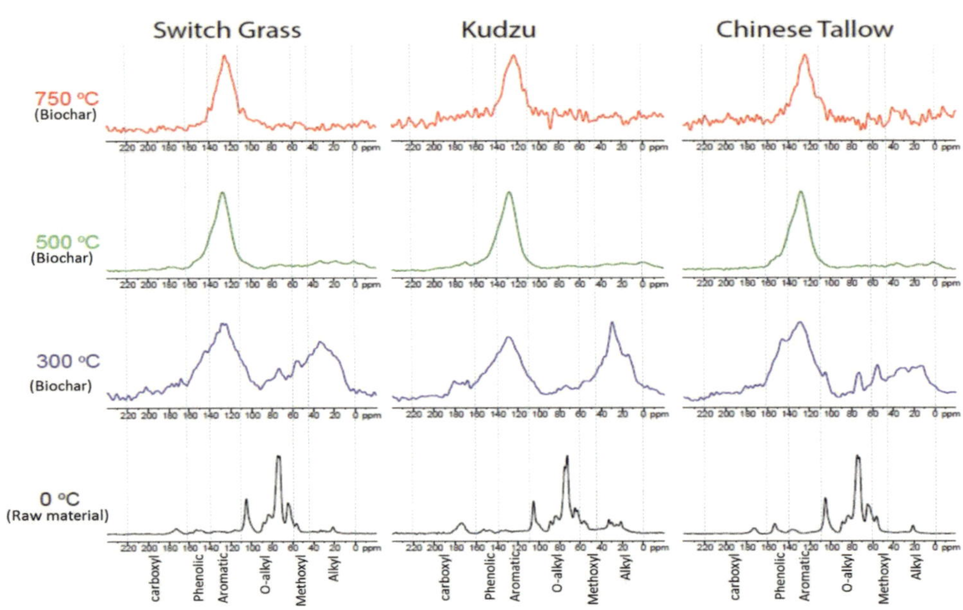

Figure 6.
Biochar carbon functional groups as determined by Nuclear Magnetic Resonance (NMR); modified from [1].

5. Biochar pH influence phosphorus sorption–desorption

The influence of the changes in pH after biochar application seems to vary between different studies. Biochar application reduces soil acidity altering P complexity with metals such as Al^{3+}, Fe^{3+} and Ca^{2+} affecting P availability, sorption and desorption (**Figure 3**) [13, 52]. As a result of higher alkalinity, biochars from legume plants increase pH much compared to biochars from non-legume plants [52].

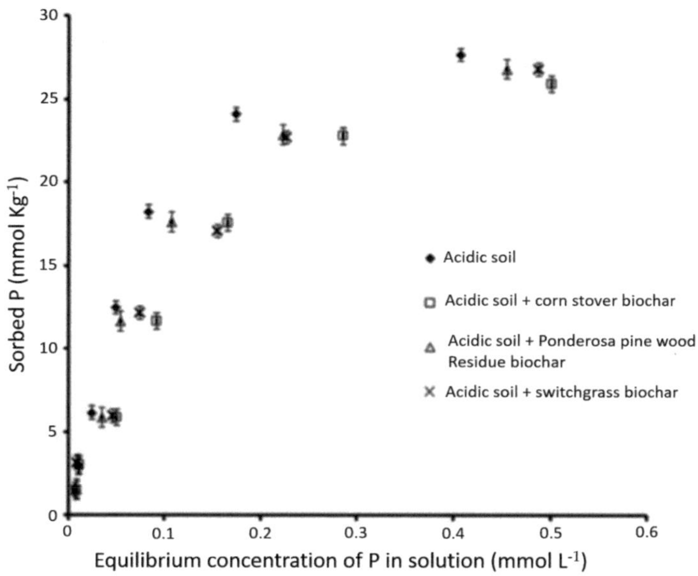

Figure 7.
P sorption by acidic soil and acidic soil (Grummit series) incubated with corn stover, Ponderosa pine wood residue or switchgrass biochars at 4 g/kg soil for 30 days. Each data point is the mean of four replications with standard error. Modified from [11].

Biochar has potential to adsorb cations such as Al^{3+}, Fe^{3+} and Ca^{2+}, which can lead to delayed P adsorption or precipitation; in addition organic molecules could sorb onto biochar reducing its capacity to chelate with Al^{3+}, Fe^{3+} and Ca^{2+} in soil (**Figure 7**) [53]. Ref. [53] reported that P sorption was increased in acidic soil but decreased in alkaline soil, and attributed increased P sorption after biochar addition to Ca induced P sorption/precipitation and also reported that P sorption was less affected by Fe and Al oxides. In contrast, incorporation of 4% biochar into acidic soil reduced the P sorption and increased availability of sorbed P, whereas application of alkaline biochars to calcareous soil increased P sorption significantly and decreased availability of sorbed P [11]. The increase in soil pH reportedly enhances the solubility of P [54] but, in contrast [1], demonstrated that addition of biochar with high pH, Ca concentration and aromatic C reduced P solubility (**Figures 3–5**). Ref. [55] indicated that colloidal and nano-sized MgO (periclase) particles on the biochar surface played the key role in providing adsorption sites for aqueous phosphate [55]. In addition, initial solution pH and coexisting anions have potential to affect adsorption of P on biochar.

6. Feedstock/biomass type and pyrolysis temperature influence phosphorus sorption

There have been diverse findings on the effect of biochar on P sorption, and in some studies, biochar application promoted availability and uptake of P as a result of biochar high anion exchange capacity; reduction of availability of Al and Fe in soil resulted in less P fixation (**Figure 8**) [50, 56]. Biochar feedstock and pyrolysis temperature affect its chemical composition and surface characteristics influencing biochar P sorption and desorption capacity [1, 11, 13]. Biochar produced from different biomass often has very different chemical characteristics and as a result influences the P sorption capacity of biochar [1, 11]. In [1], it is reported that corn stover biochar (79%) had greater sorption followed by switchgrass biochar (76%),

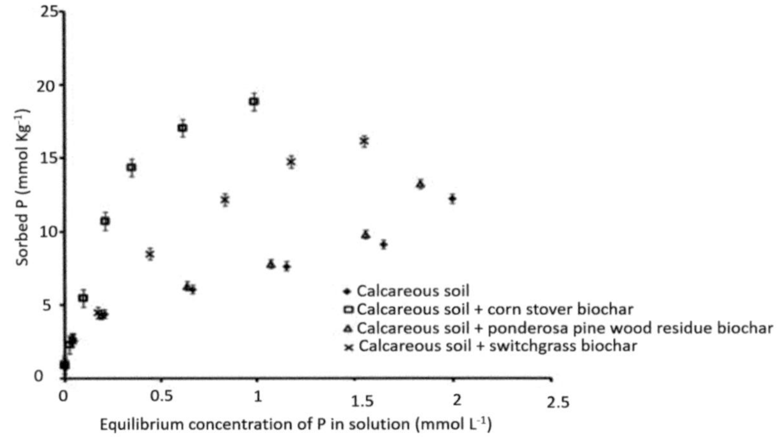

Figure 8.
P sorption calcareous soil (Langhei series) and calcareous soil incubated with corn stover, Ponderosa pine wood residue, and switchgrass biochars at 40 g/kg soil for 30 days. Each data point is the mean of four replications with standard error. Modified from [11].

Temperature	200°C	300°C	400°C	500°C	550°C	650°C	750°C	*P*-value
Switchgrass								
BC recovery (%)	78.0 ± 1.5^{a}	46.2 ± 1.1^{b}	31.7 ± 1.3^{c}	35.4 ± 2.0^{c}	24.9 ± 0.3^{d}	24.1 ± 0.2^{d}	22.8 ± 0.2^{d}	<0.0001
pH	7.1 ± 0.4^{b}	8.0 ± 0.3^{b}	9.3 ± 1.0^{ab}	9.0 ± 0.7^{ab}	10.7 ± 0.04^{a}	11.3 ± 0.1^{a}	11.3 ± 0.1^{a}	0.0001
Kudzu								
BC recovery (%)	74.0 ± 2.3^{a}	38.8 ± 0.2^{b}	29.8 ± 2.0^{c}	26.0 ± 1.1^{cd}	26.9 ± 0.5^{cd}	24.6 ± 0.1^{cd}	23.4 ± 0.2^{d}	<0.0001
pH	7.8 ± 0.03^{e}	9.5 ± 0.1^{d}	10.5 ± 0.03^{c}	11.3 ± 0.1^{b}	11.1±0.2b	12.0 ± 0.03^{a}	12.4 ± 0.01^{a}	<0.0001
Chinese Tallow								
BC recovery (%)	80.0 ± 0.4^{a}	36.5 ± 2.6^{b}	23.7 ± 0.3^{c}	22.2 ± 0.3^{cd}	20.6 ± 0.9^{cd}	20.8 ± 0.8^{cd}	17.5 ± 0.6^{d}	<0.0001
pH	7.0 ± 0.03^{c}	7.0 ± 0.9^{c}	8.8 ± 0.1^{b}	8.9 ± 0.05^{b}	8.9 ± 0.1^{b}	10.2 ± 0.1^{ab}	11.2 ± 0.3^{a}	<0.0001

Data indicate mean ± SEM (standard error of mean) between temperatures. Different letters superscript along the row indicate Tukey HSD significant difference between means of biochar recovery and pH under different temperatures. Modified from [1].

Table 1.
Biochar recovery and pH.

while Ponderosa pine wood residue biochar (31%) exhibited the lowest biochar sorption. Increasing biochar pyrolysis temperature leads to loss of the volatile compounds and increased aromatic carbon, elevated pH, biochar yield decreases (**Table 1**), increased surface area and decreased surface functional groups that provide exchange capacity (**Figures 4** and **5**) [1, 14]. Beet tailing biochar pyrolyzed at 600°C adsorbed P; the adsorption was suggested to have occurred due to binding sites contained in colloidal and nano-sized MGO particles on the biochar surface [55]. [57] indicated that similar to P sorption, desorption is also influenced by feedstock, for example Inga exhibited less sorption of P compared to Lacre and Embauba feedstocks but also desorbed less P (**Figures 9** and **10**) [57].

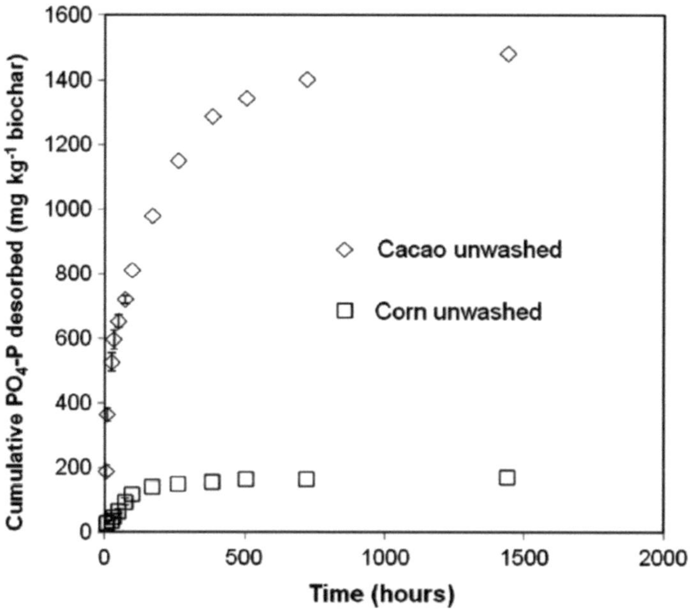

Figure 9.
Desorption of PO^4–P from washed biochars plotted as the cumulative fraction desorbed. Modified from [58].

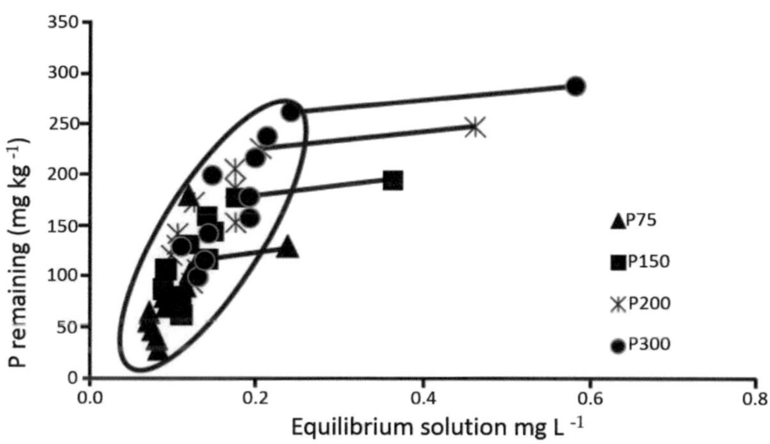

Figure 10.
Phosphorus desorption curve data for the sequential desorption of P from a degraded tropical soil with biochar added. The points represent desorption of P from the soil/biochar complex after the addition of 75, 150, 200 and 300 mg P/kg. The encircled area represents a uniform desorption curve common to all treatments with a Kd between solution concentrations of 0.1 and 0.2 mg/L of approximately 1230 L/kg. Modified from [57].

Therefore, P sorption–desorption varies between feedstock type and pyrolysis temperature levels, providing a possibility of designing biochars for specific soil management objectives (**Figures 4**, 5 and **10**) [57].

7. Biochar potential to mitigate phosphorus eutrophication

Biochar has potential to mitigate P eutrophication in the water bodies through its application in the farm. However, it is clear that biochar P sorption capacity is feedstock specific and is highly influenced by pyrolysis temperature [1]. Therefore, it is important to understand the interaction between feedstock type and pyrolysis temperature and how they influence biochar acidity/alkalinity, surface area, aromatic carbon, cation exchange capacity, surface charge density, biochar internal porosity and presence of both polar and non-polar surfaces sites that promote nutrient absorption [1, 13, 39, 41]. Biochar that promotes P sorption rather than release of P should be considered for P eutrophication mitigation, and biochar should be designed to meet specific management practices [57]. When applying biochar on farm, it is also important to consider the erodibility of biochar to water bodies; P is easily translocated to water bodies as particulate P [32, 34]. Therefore consideration of influence of biochar on erosion is important, for example in conditions where biochar reduces soil erosion, it results in reduced particulate P losses [59].

8. Conclusion

Biochar has potential to mitigate P eutrophication in the water bodies through biochar application on farm. However, biochar exhibits broad spectrum of properties which are mainly influenced by feedstock type and pyrolysis temperature and as a result influences P sorption. Some biochars have exhibited P sorption while others exhibited P desorption. Carbon composition of biochar and pH values of biochar and soils are major drivers of P sorption. It is not appropriate to extrapolate any single study findings to any other material as a result of the wide diversity in terms of biochar influence on P sorption. Therefore every material needs to be evaluated and its potential for P sorption determined. The usefulness of biochar on P sorption is dependent on its characteristics; to mitigate P eutrophication, employment of biochar with high P sorption is important.

Acknowledgements

Special thanks to College of Agriculture and Food Sciences, Florida A&M University, for providing a conducive environment for writing this book chapter. This work was supported by USDA/NIFA (1890 Evans-Allen Research Program), USDA-Forest Service Grant Number 17-CA-11330140-027 and USDA-ARS Grant Number 58-3070-7-009.

Author details

Lucy W. Ngatia[1*], Johnny M. Grace III[2], Daniel Moriasi[3], Alejandro Bolques[1], George K. Osei[4] and Robert W. Taylor[1]

1 College of Agriculture and Food Sciences, Florida A&M University, Tallahassee, Florida, USA

2 USDA-Forest Service, Southern Research Station, Tallahassee, Florida, USA

3 USDA-ARS Grazinglands Research Laboratory, El Reno, Oklahoma, USA

4 School of Environment, Florida A&M University, Tallahassee, Florida, USA

*Address all correspondence to: lucy.ngatia@famu.edu

References

[1] Ngatia LW, Hsieh YP, Nemours D, Fu R, Taylor RW. Potential phosphorus eutrophication mitigation strategy: Biochar carbon composition, thermal stability and pH influence phosphorus sorption. Chemosphere. 2017;**180**:201-211

[2] Xu H, Paerl HW, Qin B, Zhu G, Gao G. Nitrogen and phosphorus inputs control phytoplankton growth in eutrophic Lake Taihu, China. Limnology and Oceanography. 2010;**55**(1):420-432

[3] Paerl HW. Nuisance phytoplankton blooms in coastal, estuarine, and inland waters. Limnology and Oceanography. 1988;**33**:823-847

[4] Carpenter SR, Caraco NF, Correll DL, Howarth RW, Sharpley AN, Smith VH. Nonpoint pollution of surface waters with phosphorus and nitrogen. Ecological Applications. 1998;**8**:559-568

[5] Bennett EM, Carpenter SR, Caraco NF. Human impact on erodable phosphorus and eutrophication: A global perspective: Increasing accumulation of phosphorus in soil threatens rivers, lakes, and coastal oceans with eutrophication. BioScience. 2001;**51**(3):227-234

[6] Fluck RC, Fonyo C, Flaig E. Land-use-based phosphorus balances for Lake Okeechobee, Florida, drainage basins. Applied Engineering in Agriculture. 1992;**8**:813-820

[7] Jaworski NA, Groffman PM, Keller AA, Prager JC. A watershed nitrogen and phosphorus balance: The upper Potomac River basin. Estuaries. 1992;**15**:83-95

[8] Stutter MI, Shand CA, George TS, Blackwell MSA, Bol R, MacKay RL, et al. Recovering phosphorus from soil: A root solution? Environmental Science and Technology. 2012;**46**:1977-1978

[9] Condron LM, Turner BL and Cade-Menun BJ. The chemistry and dynamics of soil organic phosphorus. In: Sims JT, Sharpley AN, editors. Phosphorus: Agriculture and the Environment. Madison, Wisconsin, USA: ASA-CSSASSSA; 2005. pp. 87-121

[10] Negassa W, Leinweber P. How does the Hedley sequential phosphorus fractionation reflect impacts of land use and management on soil phosphorus: A review. Journal of Plant Nutrition and Soil Science. 2009;**172**:305-325

[11] Chintala R, Schumacher TE, McDonald LM, Clay DE, Malo DD, Papiernik SK, et al. Phosphorus sorption and availability from biochars and soil/biochar mixtures. Clean: Soil, Air, Water. 2014;**42**(5):626-634

[12] Cornelissen G, Gustafsson O, Bucheli TD, Jonker MTO, Koelmans AA, Van Noort PCM. Extensive sorption of organic compounds to black carbon, coal, and kerogen in sediments and soils: Mechanisms and consequences for distribution, bioaccumulation, and biodegradation. Environmental Science and Technology. 2005;**39**(18):6881-6895

[13] Wang T, Arbestain MC, Hedley M, Bishop P. Predicting phosphorus bioavailability from high-ash biochars. Plant and Soil. 2012;**357**:173-187

[14] Novak JM, Lima I, Xing B, Gaskin JW, Steiner C, Das KC, et al. Characterization of designer biochar produced at different temperatures and their effects on a loamy sand. 2009;**3**:195-2006

[15] Afif E, Matar EA, Torrent J. Availability of phosphate applied to calcareous soils of West Asia and North Africa. Soil Science Society of America Journal. 1993;**57**:756-760

[16] Sample ECR, Soper RJ, Racz GJ. Reaction of phosphate fertilizers in

soils. In: Dinauer RC, editor. The Role of Phosphorus in Agriculture. Madison, WI: American Society of America; 1980. pp. 263-304

[17] Lindsay WL. Chemical Equilibria in Soils. New York: John Wiley & Sons; 1979

[18] Amer FA, Mahmoud AA, Sabet V. Zeta-potential and surface-area of calcium carbonate as related to phosphate sorption. Soil Science Society of America Journal. 1985;**49**:1137-1142

[19] Marschner H. Mineral Nutrition of Higher Plants. London: Academic Press; 1995

[20] Turner BL. Organic phosphorus transfer from terrestrial to aquatic environments. In: Turner BL, Frossard E, Baldwin DS, editors. Organic Phosphorus in the Environment. Wallingford: CAB International; 2005. pp. 269-295

[21] Tarafdar JC, Claassen N. Organic phosphorus compounds as a phosphorus source for higher plants through the activity of phosphatases produced by plant roots and microorganisms. Biology and Fertility of Soils. 1988;**5**:308-312

[22] Adams MA, Pate JS. Availability of organic and inorganic forms of phosphorus to lupins (*Lupinus* spp.). Plant and Soil. 1992;**145**:107-113

[23] Bostrom B. Potential mobility of phosphorus in different types of lake sediment. Internationale Revue der Gesamten Hydrobiologie. 1984;**69**:457-475

[24] Bostrom B, Andersen JM, Fleischer S, Jansson M. Exchange of phosphorus across the sediment water interface. Hydrobiologia. 1988;**170**:229-244

[25] Redshaw CJ, Mason CF, Hayes CR, Roberts RD. Factors influencing phosphate exchange across the sediment–water interface of eutrophic reservoirs. Hydrobiologia. 1990;**192**:233-245

[26] Perkins RG, Underwood GJC. The potential for phosphorus release across the sediment–water interface in a eutrophic reservoir dosed with ferric sulphate. Water Research. 2001, 2001;**35**(6):1399-1406

[27] Cordell D, Drangert JO, White S. The story of phosphorus: Global food security and food for thought. Global Environmental Change. 2009;**19**:292-305

[28] Bouwman L, Goldewijk KK, Van Der Hoek KW, Beusen AHW, Van Vuurena DP, Willems J, et al. Exploring global changes in nitrogen and phosphorus cycles in agriculture induced by livestock production over the 1900-2050 period. Proceedings of the National Academy of Sciences of the United States of America. 2013;**110**(52):20882-20,887

[29] Sharpley AN, Reikolainen S. Phosphorus in agriculture. In: Tunney H, Carton OT, Brookes PC, Johnston AE, editors. Phosphorus Loss from Soil to Water. New York: CAB International; 1997. pp. 7-53

[30] Daniel TC, Sharpley AN, Lemunyon JL. Agricultural phosphorus and eutrophication: A symposium overview. Journal of Environmental Quality. 1998;**27**:251-257

[31] Daniel TC, Sharpley AN, Edwards DR, Wedepohl R, Lemunyon JL. Minimizing surface water eutrophication from agriculture by phosphorus management. Journal of Soil and Water Conservation. 1994;**49**:30-38

[32] Sharpley AN, Chapra SC, Wedepohl R, Sims JT, Daniel TC, Reddy KR. Managing agricultural phosphorus for protection of surface waters: Issues and options. Journal of Environmental Quality. 1994;**23**:437-451

[33] Haygarth PM, Condron LM, Heathwaite AL, Turner BL, Harris GP. The phosphorus transfer continuum: Linking source to impact with an interdisciplinary and multi-scaled approach. Science of the Total Environment. 2005;**344**:5-14

[34] Meinikmann K, Hupfer M, Lewandowski J. Phosphorus in groundwater discharge—A potential source for Lake Eutrophication. Journal of Hydrology. 2015;**524**:214-226

[35] Antal MJ, Gronli M. The art, science, and technology of charcoal production. Industrial and Engineering Chemistry Research. 2003;**42**:1619-1640

[36] Benaddi H, Bandosz TJ, Jagiello J, Schwarz JA, Rouzaud JN, Legras D, et al. Surface functionality and porosity of activated carbons obtained from chemical activation of wood. Carbon. 2000;**38**:669-674

[37] Guo Y, Rockstraw DA. Physicochemical properties of carbons prepared from pecan shell by phosphoric acid activation. Bioresource Technology. 2007;**98**:1513-1521

[38] Strelko V Jr, Malik DJ, Streat M. Characterization of the surface of oxidized carbon adsorbents. Carbon. 2002;**40**:95-104

[39] Liang B, Lehmann J, Solomon D, Kinyangi J, Grossman J, O'Neill B, et al. Black carbon increases cation exchange capacity in soils. Soil Science Society of America Journal. 2006;**70**:1719-1730

[40] Cheng CH, Lehmann J, Engelhard MH. Natural oxidation of black carbon in soils: Changes in molecular form and surface charge along aclimosequence. Geochimica et Cosmochimica Acta. 2008;**72**:1598-1610

[41] Laird D, Fleming P, Wang BQ, Horton R, Karlen D. Biochar impact on nutrient leaching from a Midwestern agricultural soil. Geoderma. 2010;**158**:436-442

[42] Knowles OA, Robinson BH, Contangelo A, Clucas L. Biochar for the mitigation of nitrate leaching from soil amended with biosolids. Science of the Total Environment. 2011;**409**:3206-3210

[43] Spokas KA, Novak JM, Venterea RT. Biochar's role as an alternative N-fertilizer: Ammonia capture. Plant and Soil. 2012;**350**:35-42

[44] Taghizadeh-Toosi A, Clough TJ, Sherlock RR, Condron LM. Biochar adsorbed ammonia is bioavailable. Plant and Soil. 2012;**350**:57-69

[45] Soinne H, Hovi J, Tammeorg P, Turtola E. Effect of biochar on phosphorus sorption and clay soil aggregate stability. Geoderma. 2014;**219-220**:162-167

[46] Hollister CC, Bisogni JJ, Lehmann J. Ammonium, nitrate, and phosphate sorption to and solute leaching from biochars prepared from corn stover (*Zea mays* L.) and oak wood (*Quercus* spp.). Journal of Environmental Quality. 2013;**42**:137-144

[47] Yao Y, Gao B, Zhang M, Inyang M, Zimmerman AR. Effect of biochar amendment on sorption and leaching of nitrate, ammonium, and phosphate in a sandy soil. Chemosphere. 2012;**89**:1467-1471

[48] Nelson NO, Agudelo SC, Yuan WQ, Gan J. Nitrogen and phosphorus availability in biochar-amended soils. Soil Science. 2011;**176**:218-226

[49] Beck DA, Johnson GR, Spolek GA. Amending green roof soil with biochar to affect runoff water quantity and quality. Environmental Pollution. 2011;**159**:2111-2118

[50] Novak JM, Busscher WJ, Laird DL, Ahmedna M, Watts DW, Niandou AS.

Impact of biochar amendment on fertility of a southeastern coastal plain soil. Soil Science. 2009;**174**:105-112

[51] Chen BL, Yuan MX. Enhanced sorption of polycyclic aromatic hydrocarbons by soil amended with biochar. Journal of Soils and Sediments. 2011;**11**:62-71

[52] Yuan JH, Xu RK, Wang N, Li JY. Amendment of acid soils with crop residues and biochars. Pedosphere. 2011;**21**:302-308

[53] Xu G, Sun J, Shao H, Chang SX. Biochar had effects on phosphorus sorption and desorption in threesoils with differing acidity. Ecological Engineering. 2014;**62**:54-60

[54] Hartikainen H, Yli-Halla M. Solubility of soil phosphorus as influenced by urea. Zeitschrift für Pflanzenernährung und Bodenkunde. 1996;**159**:327-332

[55] Yao Y, Gao B, Inyang M, Zimmerman AR, Cao XD, Pullammanappallil P, et al. Removal of phosphate from aqueous solution by biochar derived from anaerobically digested sugar beet tailings. Journal of Hazardous Materials. 2011;**190**:501 507

[56] Deluca TH, Mackenzie MD, Gundale MJ. Biochar effects on soil nutrient transformation. In: Lehmann J, Joseph S, editors. Biochar for Environmental Management Science and Technology. Chapter 14. London: Earthscan; 2009. pp. 251-280

[57] Morales MM, Comerford N, Guerrini IA, Falcao NPS, Reeves J, Reeves JB. Sorption and desorption of phosphate on biochar and biochar–soil mixtures. Soil Use and Management. 2013;**29**:306-314

[58] Hale SE, Alling V, Martisen V, Mulder J, Breedveld GD, Coenelissen G. The sorption and desorption of phosphate-P, ammonium-N and nitrate-N in cacao shell and corn cob biochar. Chemosphere. 2013;**91**:1612-1619

[59] Jien S, Wang C. Effects of biochar on soil properties and erosion potential in a highly weathered soil. CATENA. 2013;**110**:225-223